LIFE CHEMISTRY RESEARCH

Biological Systems

LIFE CHEMISTRY RESEARCH

Biological Systems

Edited by
**Roman Joswik, PhD, Gennady E. Zaikov, DSc, and
A. K. Haghi, PhD**

Apple Academic Press Inc. | Apple Academic Press Inc.
3333 Mistwell Crescent | 9 Spinnaker Way
Oakville, ON L6L 0A2 | Waretown, NJ 08758
Canada | USA

Exclusive worldwide distribution by CRC Press, a member of Taylor & Francis Group
No claim to original U.S. Government works

ISBN 13: 978-1-77463-084-6 (pbk)
ISBN 13: 978-1-77188-068-8 (hbk)

Library and Archives Canada Cataloguing in Publication

Life chemistry research : biological systems / edited by Roman Joswik, PhD, Gennady E. Zaikov, DSc, and A.K. Haghi, PhD.

Includes bibliographical references and index.
ISBN 978-1-77188-068-8 (bound)
1. Chemistry. 2. Biochemistry. 3. Biological systems. I. Haghi, A. K., editor II. Zaikov, G. E. (Gennadiĭ Efremovich), 1935-, author, editor III. Joswik, Roman, editor

QD31.3.L53 2015   540   C2015-902234-7

Library of Congress Cataloging-in-Publication Data

Life chemistry research : biological systems / Roman Joswik, PhD, Gennady E. Zaikov, DSc, and A.K. Haghi, PhD, editors.

pages cm
Includes bibliographical references and index.
ISBN 978-1-77188-068-8 (alk. paper)
1. Bioremediation. 2. Biology. 3. Chemistry, Organic. I. Joswik, Roman, editor. II. Zaikov, G. E. (Gennadii Efremovich), 1935- editor. III. Haghi, A. K., editor.

TD192.5.L54 2015   572--dc23   2015011054

ABOUT THE EDITORS

Roman Joswik, PhD

Roman Joswik, PhD, is Director of the Military Institute of Chemistry and Radiometry in Warsaw, Poland. He is a specialist in the field of physical chemistry, chemical physics, radiochemistry, organic chemistry, and applied chemistry. He has published several hundred original scientific papers as well as reviews in the field of radiochemistry and applied chemistry.

Gennady E. Zaikov, DSc

Gennady E. Zaikov, DSc, is Head of the Polymer Division at the N. M. Emanuel Institute of Biochemical Physics, Russian Academy of Sciences, Moscow, Russia, and Professor at Moscow State Academy of Fine Chemical Technology, Russia, as well as Professor at Kazan National Research Technological University, Kazan, Russia. He is also a prolific author, researcher, and lecturer. He has received several awards for his work, including the Russian Federation Scholarship for Outstanding Scientists. He has been a member of many professional organizations and on the editorial boards of many international science journals.

A. K. Haghi, PhD

A. K. Haghi, PhD, holds a BSc in urban and environmental engineering from University of North Carolina (USA); a MSc in mechanical engineering from North Carolina A&T State University (USA); a DEA in applied mechanics, acoustics and materials from the Université de Technologie de Compiègne (France); and a PhD in engineering sciences from the Université de Franche-Comté (France). He is the author and editor of 165 books as well as 1,000 published papers in various journals and conference proceedings. Dr. Haghi has received several grants, consulted for a number of major corporations, and is a frequent speaker to national and international audiences. Since 1983, he served as a professor at several universities. He is currently Editor-in-Chief of the *International Journal of Chemoinformatics and Chemical Engineering and Polymers Research Journal* and on the editorial boards of many international journals. He is a member of the Canadian Research and Development Center of Sciences and Cultures (CRDCSC), Montreal, Quebec, Canada.

CONTENTS

LIST OF CONTRIBUTORS

O. M. Alekseeva
Emanuel Institute of Biochemical Physics, Russian Academy of Sciences, Moscow, ul. Kosygina, 4, Moscow, 119334; E-mail: olgavek@yandex.ru

A. I. Bastrakov
A.N. Severtsov Institute of Ecology and Evolution, RAS, 33 Leninskij prosp., Moscow, 119071, Russia.

O. S. Berdyugina
FGBU "I.I. Mechnikov Research Institute for Vaccines and Sera" RAMS., Moscow 105064 Maliy Kazenniy per. 5a. E-mail labpitsred@yandex.ru

A. I. Beresnev
Institute of Microbiology, National Academy of Sciences, 220141, Minsk, Belarus.

Sanjay Kumar Bharti
School of Pharmaceutical Sciences, Guru Ghasidas Vishwavidyalaya (A Central University), Bilaspur, Chattisgarh, India

L. P. Blinkova
FGBU "I. I. Mechnikov Research Institute for Vaccines and Sera" RAMS. Moscow 105064 Maliy Kazenniy per. 5a. e-mail labpitsred@yandex.ru.

S. B. Bokieva
N. N. Semenov Institute of Chemical Physics, RAS, Moscow, Russia

A. A. Brilliant
GBUZSO Institute of Medical Cell Technologies, 620036 Yekaterinburg.

E. B. Burlakova
Emanuel Institute of Biochemical Physics RAS, 119334, Moscow, Russia

M. S. Chirikova
Institute of Microbiology, National Academy of Sciences, Belarus, 220141, Minsk, Belarus, E-mail: margarita.chirikova@mail.ru.

O. V. Dmitrieva
FGBU "I. I. Mechnikov Research Institute for Vaccines and Sera" RAMS., Moscow 105064 Maliy Kazenniy per. 5a. e-mail labpitsred@yandex.ru.

Sergey Gaydamaka
Moscow State University, Chemistry Faculty, Department of Chemical Enzymology. 119991, Moscow, Russia.

N. Yu. Gerasimov
Emanuel Institute of Biochemical Physics RAS, 119334, Moscow, Russia

M. D. Goldfein
Saratov State University named after N.G. Chernyshevsky, Russia, goldfeinmd@mail.ru

A. N. Goloshchapov
Emanuel Institute of Biochemical Physics RAS, 119334, Moscow, Russia

N. A. Grebenkina
Higher Chemical College, RAS, Moscow, Russia

K. Z. Gumargalieva
N. N. Semenov Institute of Chemical Physics, RAS, Moscow, Russia

A. N. Inozemtsev
M. V. Lomonosov MSU, Biological Faculty, Leninskie Gory, 119991, Moscow, Russia

V. V. Kasparov
Emanuel Institute of Biochemical Physics RAS, 119334, Moscow, Russia

Yu. A. Kim
Institute of Cell Biophysics, Russian Academy of Sciences, Pushchino, Moscow, Russia

M. A. Klyuchnikova
A. N. Severtzov Institute of Ecology and Evolution, 33 Leninski prospect, Moscow, 119071, Russia; e-mail: veravoznessenskaya@gmail.com

Sergei S. Kolesov
The Institute of Organic Chemistry of the Ufa Scientific Centre the Russian Academy of Science, Russia, Republic of Bashkortostan, Ufa, 450054, October Prospect 71.

A. L. Kovarskij
Emanuel Institute of Biochemical Physics RAS, 119334, Moscow, Russia.

A. A. Kozlova
A. N. Severtsov Institute of Ecology and Evolution, RAS, 33 Leninskij prosp., Moscow, 119071, Russia.

L. Z. Kravtsova
The "NTC BIO", LLC, 309292, Russia, Belgorod Region, Shebekino town, e-mail: ntcbio@mail.ru

E. I. Kulish
Bashkir State University, Russia, Republic of Bashkortostan, Ufa, 450074, ul. Zaki Validi.

S. V. Kvach
Institute of Microbiology, National Academy of Sciences, 220141, Minsk, Belarus.

C. A. Liman
the "Agroakademia", LLC, 309290 Russia, Belgorod region, Shebekino town, A., e-mail: agroakademia@mail.ru

Murygina Lomonosov
Moscow State University, Chemistry Faculty, Department of Chemical Enzymology. 119991, Moscow, Leninsky gory 1/11

Debarshi Kar Mahapatra
School of Pharmaceutical Sciences, Guru Ghasidas Vishwavidyalaya (A Central University), Bilaspur, Chattisgarh, India

T. V. Malanina
A.N.Severtzov Institute of Ecology & Evolution, 33 Leninski prospect, Moscow, 119071, Russia, email: veravoznessenskaya@gmail.com.

E. I. Martirosova
Emanuel Institute of Biochemical Physics, RAS, 119334, Moscow, Russia. Email: ms_martins@mail.ru

O. V. Nevrova
Emanuel Institute of Biochemical Physics RAS, 119334, Moscow, Russia

I. R. Oviya
Department of Bioinformatics, Bharathiar University, Coimbatore, India.

Yu. D. Pakhomov
FGBU "I.I. Mechnikov Research Institute for Vaccines and Sera" RAMS., Moscow 105064 Maliy Kazenniy per. 5a. e-mail labpitsred@yandex.ru.

D. S. Pavlov
A.N. Severtsov Institute of Ecology and Evolution, Russian Academy of Sciences, 119071 Russia, Moscow.

I. G. Plashchina
Emanuel Institute of Biochemical Physics, RAS, 119334, Moscow, Russia.

V. V. Podmasteryev
N.M. Emanuel Institute of Biochemical Physics, Russian Academy of Sciences, 119334, Moscow, Russia

S. V. Ponomarev
the "Bioaquapark" Innovation Centre– The Scientific Centre of the Aqua-Culture at the ASTU, 414025, Astrakhan, e-mail: kafavb@yandex.ru

V. G. Pravdin
The "NTC BIO", LLC, 309292, Russia, Belgorod region, Shebekino town, e-mail: ntcbio@mail.ru

S. D. Razumovsky
N.M. Emanuel Institute of Biochemical Physics, Russian Academy of Sciences, Mascow, Russia.

E. I. Rodionova
A. A. Kharkevich Institute for Information Transmission, 19 B. Karetny, Moscow, 127994, Russia.

E. G. Rozantsev
Saratov State University named after N.G. Chernyshevsky, Russia.

A. S. Samsonova
Institute of Microbiology, National Academy of Sciences, Belarus, 220141, Minsk, Belarus

R. Sathishkumar
Dept. of Biotechnology, Salem Sowdeswari College, Salem, India.

I. P. Savchenkova
All Russian State Research Institute of Experimental Veterinary Medicine of Ya.R. Kovalenko, 109428, Russia, Moscow, E–mail: s-ip@mail.ru

S. V. Sazonov
GBUZSO Institute of Medical Cell Technologies, 620036 Yekaterinburg.

T. P. Shakun
Institute of Microbiology, National Academy of Sciences, Belarus, 220141, Minsk, Belarus, E-mail: margarita.chirikova@mail.ru.

M. Sharanya
Department of Bioinformatics, Bharathiar University, Coimbatore, India.

Angela S. Shurshina
Bashkir State University, Russia, Republic of Bashkortostan, Ufa, 450074.

V. S. Sibirtsev
GiproRjibFlot (Research and design institute on development and exploitation of a fish fleet), lab. Technical microbiology; Instrumentalnaja ul. 8, 197022 Russia, e-mail: vs1969r@mail.ru; site: http:\\www.vs1969r.narod.ru\publen.htm

Anamika Singh
Maitreyi Collage, University of Delhi, India. E-mail: 10rsingh@gmail.com

Rajeev Singh
Division of Reproductive and Child Health, Indian Council of Medical Research, New Delhi

G. G. Sivets
Institute of Bioorganic Chemistry, National Academy of Sciences, 220141, Minsk, Belarus

N. N. Skorlupkina
FGBU "I. I. Mechnikov Research Institute for Vaccines and Sera" RAMS., Moscow 105064 Maliy Kazenniy per. 5a. e-mail labpitsred@yandex.ru tel.: +7 495 916-11-52, fax: +7 495 917-54-60.

N. A. Ushakova
A. N. Severtsov Institute of Ecology and Evolution, Russian Academy of Sciences, 33 Leninskij prosp., Moscow, 119071 Russia, fax (8495) 954-55-34, e-mail naushakova@gmail.com

P. Valentina
Moscow State University, Chemistry Faculty, Department of Chemical Enzymology, 119991, Moscow, Russia.

A. E. Voznesenskaya
A. A. Kharkevich Institute for Information Transmission, 127994, Moscow, Russia

V. V. Voznessenskaya
A. N. Severtzov Institute of Ecology & Evolution, 33 Leninski prospect, Moscow, 119071, Russia, Email: veravoznessenskaya@gmail.com.

Bataeva Yulia
Federal State Budget Educational Institution of Higher Professional Education, Astrakhan State University. E-mail: aveatab@mail.ru

A. A. Zagorinsky
A. N. Severtsov Institute of Ecology and Evolution, RAS, 33 Leninskij prosp., Moscow, 119071, Russia.

G. E. Zaikov
N. M. Emanuel Institute of Biochemical Physics, Russian Academy of Sciences, Moscow 119334, Russia, Chembio@sky.chph.ras.ru

Y. M. Zasadkevich
GBUZSO Institute of Medical Cell Technologies, 620036 Yekaterinburg, Russia.

A. I. Zinchenko
Institute of Microbiology, National Academy of Sciences, 220141, Minsk, Belarus.

LIST OF ABBREVIATIONS

AD	Alzheimer's disease
AFO	ankle foot orthotics
AHB	alkylhydroxybenzenes
AMS	amikacin sulfate
AMS	antibiotics - amikacin
BSA	bovine serum albumin
CBC	cyano-bacterial communities
CBT	cognitive behavioral therapy
ChTA	chitosan acetate
CMC	critical micelle concentration
CTD	common technical document
EMEA	evaluation of medicinal products
EPR	electron paramagnetic resonance
ETP	electron transport particles
FFA	free fatty acids
FSR	fragmented sarcoplasmic reticulum
HAS	human serum albumin
INNs	international non-proprietary names
LMWC	low molecular weight chitosans
LT	longitudinal tubules
MFC	minimal fungicidal concentration
MFD	minimum fungicidal dilution
MIC	minimal inhibitory concentration
MID	minimum inhibitory dilution
NMR	nuclear magnetic resonance
PD	Parkinson's disease
PRCA	pure red cell aplasia
PRET	progressive resisted exercise training
RyR	ryanodine receptor
SIRS	system inflammatory response syndrome
SR	sarcoplasmic reticulum
TC	terminal cysternaes
TCM	traditional chinese medicine
TGA	therapeutic goods administration
UPSIT	University of Pennsylvania Smell Identification Test

LIST OF SYMBOLS

m_∞	relative amount of water in equilibrium swelling film sample
k	constant connected with parameters of interaction polymer
n	indicator characterizing the mechanism
c_0	surface concentration
τ	constant of proportionality
τ_c	rotational diffusion correlation time
V	volume of the radical
η	dynamic viscosity of the medium
K	Boltzmann constant
T	absolute temperature
n	number of cracks
l	length of a crack
h	depth
D	optical density
α	solubility coefficient

PREFACE

This book, with contributions from many world leaders in the field, is equally appropriate for graduate or research courses in biochemistry. The book has been extensively class-tested and includes tutorials in biology and biochemistry to aid students of varying backgrounds. This exciting new book will be a must-read for years to come for all students and researchers interested in the field of biological chemistry.

This volume also contains experiments related to the content of biological chemistry courses as well as basic/preparatory chemistry courses. These research studies give students an opportunity to go beyond the lectures and words in the textbook to experience the scientific process from which conclusions and theories are drawn.

This book:
- Focuses on fundamental and relevant connections between chemistry and life.
- Elegantly portrays the complementary nature of chemistry and biology. By describing biological processes in detailed chemical terms, the authors have provided a resource that provides an unparalleled look into the fascinating and emerging field of chemical biology.
- Satisfies a major need in chemistry curricula, bridging the gap between introductory organic chemistry and biochemistry/biology.
- Delivers need-to-know information in a succinct style for today's students.

PART I
BIOLOGICAL MEDICINE

CHAPTER 1

A STEP TOWARD PERSONALIZED MEDICINE

I. R. OVIYA

Department of Bioinformatics, Bharathiar University, Coimbatore - 641 046, India

CONTENTS

1.1   INTRODUCTION

The age-old practice of traditional medicine is gaining momentum today. The traditional and alternative medicinal practices like Ayurveda, Siddha, Unani, Kampo, traditional Chinese medicine, etc., have promising therapeutic values. They constitute the group of evidence-based medicine. Though not scientifically proven then, but the practical knowledge of our forefathers have unbelievably exceeded today's scientific practice when the whole foundation was based upon personalized medicinal concept or "Prakriti" which is also mentioned in our ancient text as follows [3], [2]:

> *"Every individual is different from another*
>
> *and hence should be considered as a different entity.*
>
> *As many variations are there in the Universe, all are seen in Human being".*

With the advancement in biological science, the root cause of all human diseases is mapped to the genomic and phenotypic variations which are important to understand for personalized medicine concept. Prakriti-based medicine is based on three doshas, namely, vata, pitta, and kapha explaining the different conditions of the body.

1.2   WHAT IS ALTERNATIVE MEDICINE?

Alternative medicine which includes Ayurveda, Siddha, Unani, Kampo, traditional Chinese medicine, to name a few, is a substitute for present conventional medicine, namely Allopathy. They are practiced with traditional knowledge for healing in different parts of the world. The ancestral medicinal knowledge documented by various scholars of their era is now being used to gain knowledge for alternative medicinal practice.

1.2.1   PROS AND CONS

1.   Alternative medicines are based on evidence-based practice since ancient times. Thus, it has a strong foundation for sustaining its practice.
2.   They are believed to possess lesser side effects.
3.   Alternative medicine is based on personalized medicine concept. Hence, drug selection and dosage quantity is based on the prakriti or tridosha of a person under treatment.
4.   Alternative medicines are believed to cure some of the most menacing disease of present era which the present conventional medicine is unable to cure.

5. Presently there is no standardization of herbal medicine which compromises on the purity of the medicine.
6. Still various practices of the tribal communities are undisclosed. Hence, very little is known of different traditional remedies practiced today.
7. Scientific research is still lagging in understanding the mechanism of action of the herbal formulations and its synergistic activity in the biological system.

1.3 WHAT IS BIOINFORMATICS?

Bioinformatics is an interdisciplinary area for analyzing the biological systems with a computational approach. The area of bioinformatics has diversified into many sister areas of research namely, genomics, proteomics, metabolomics, systems biology, pharmacogenomics, etc. More recently few new areas seem to emerge like ayugenomics [2], reverse pharmacogenomics [1], and pathogenomics [11]. Few areas are briefly described here:

1.3.1 GENOMICS

Genomics is the study of sequence, structure, and function of genome. The completion of the human genome project held high hopes of the scientists in undeciphering the genetic puzzle in humans in medical perspective. Breakthrough in establishing basis of individual differences with VNTRs, SNPs, EST, and other genomic elements gave hope toward a personalized medication approach. However, the science of human body is too complex to understand and apply these concepts to the whole.

1.3.2 PROTEOMICS

Proteomics is the study of structure and function of proteins. In recent times, there has been a boom in structure elucidation of various proteins and its mode of action. These structures give us better understanding of the protein behavior with different molecules. The identification of the active site and the active residues are the main components into probing the possible drug-target interaction.

1.3.3 TRANSCRIPTOMICS

Transcriptomics is the study of expression level of different proteins in a given cell population. It is also referred as expression profiling and is based on microarray technology. This technology is used to study the differences in the expression of different genetic elements in more than one condition such as normal vs. disease conditions.

1.3.4   METABOLOMICS

The study of various intermediate metabolites in cellular processes and decoding the chemical fingerprints is known as metabolomics.

1.3.5   SYSTEMS BIOLOGY

Systems biology focuses on integrating the data from genomics, proteomics, transcriptomics, and metabolomics. It focuses on computational model building using the data involving metabolic networks or cell signaling networks. Network building is an integral part of systems biology.

1.3.6   PHARMACOGENOMICS

Pharmacogenomics aims toward personalized medicine approach. It involves technology combining genomics, proteomics, metabolomics, etc., to analyze the individual's genetic makeup affecting his drug response.

This integrative field is still in its infancy with respect to understanding of complexity of the biological entities. However, the advances made in the sister areas can't be ignored.

1.4   BIOINFORMATICS AND ALTERNATIVE MEDICINE—THE PRESENT SCENARIO

The promises of the computational models for studying the biological systems is not new but the advent of the technology post the completion of the human genome project has been huge. High throughput sequencing data, microarray, and expression data compiled systematically in a database is a boon for the modern biologist for extracting the experimental knowledge and its application henceforth. The present-day approaches in alternative medicine are mainly focuses on QSAR studies, molecular modeling, docking, dynamic, and other validation studies. These studies are mainly focused on the application of the developed strategies. The prospects of bioinformatics is not only limited to the usage of available data and softwares but to develop more robust network models to comply with the dynamicity of the biological systems. Human system is complex as well as variable. The earliest scientific approach was restricted to understand the human complexicity. Recent studies have to be extended to deal with the variability in genome induced due to physiological and biological changes. The computational biology plays an integral role in bioinformatics in developing the softwares for analysis. The efficiency of the computational techniques is dependent on the mathematical models implemented using programming languages which can deal with both complexity and variability as mentioned.

1.4.1 DATABASES, TEXT MINING METHODS, OR LITERATURE-BASED STUDIES

The information on the traditional medicine is scattered and the present-day need is to gather them and present it to the researchers in a more organized manner. Various kind of information are available in different databases which minimize our searing work and allow us to effective utilize the information for further analysis. The traditional practices and usages that have been documented in different books, scriptures, and sometimes they have not been documented but have been in practice with the group or tribe for a long time. Such kind of information is highly valuable and cannot be harnessed at one place. Hence the research work comprising of field visit, collection of traditional information, and documentation in modern form (databases) is important to give other researchers a peep into the data and to carry out the further pharmacognostical or phytochemical work. Medicinal pharmacogenomics is based in the information available likewise. Similarly, along with documentation, web search needs to be more robust and specific. Text mining approaches improves not only efficiency of search but also saves time. Few databases for the reference of the reader are mentioned below:

Databases	Information	References
Supernatural	Available natural compounds	[4]
TCM-ID	Herbal ingredient, structure, function, therapeutic details	[22]
TCMGeneDIT	Association information about TCM	[5]
HIT	Herbs and their protein target information	Hao et al. (2011)
SWEETLEAD	Approved drugs, regulated chemicals, herbal isolates for CADD	[18]
INpacdb	Indian plant anticancer compounds	[20]
NAPROC-13	Indian medicinal plants	[17]

1.4.2 EXPRESSION STUDIES

Microarray analysis, spectroscopic methods, and other high throughput technology have led to the increase in the pharmacological understanding of the disease conditions. The differential gene, protein expression can help us to identify the candidate gene and thus help to identify the drug target. The expression data can be accessed by various public database like GEO, ArrayExpress, etc.

1.4.3 TRADITIONAL CHINESE MEDICINE (TCM)

The worldwide acceptance of TCM makes it a promising source of experimental studies. The concept of personalized medicine is not new and it is being practiced for thousands of years. The basis of TCM practice lies in Zheng (i.e., personalized identification and classification of the symptoms and treating it individually). Zheng describes the overall physical status of the human body based on the genotypic and phenotypic differences arising due to heredity, SNPs, environmental variations, and nutrition wherein the patients can classified into characteristic hot or cold Zheng. This has given our modern day researchers to probe into the effectiveness of TCM and its underlying mechanism. Various research papers suggest *in-silico* approach to solve the puzzle of TCM. The sequencing data, text mining resources, pharmacogenomics, microarray expression analysis, etc., all gave a reason to systematically investigate TCM further [19]. More recently systems biology and network pharmacology is the principle choice of carrying out study designs on TCM. Network pharmacology integrates the available chemical and metabolomic knowledge derived from experimental results for drug-target interaction or network studies in a computer simulated model mimicking the living systems. The mode of actions and the scientific basis behind TCM's principle is still unclear. There is a curiosity to unveil the principles behind TCM life systems. Kang et al. developed novel entropy-based models like acquired life entropy, acquired life entropy flow, and acquired life entropy production using the experimental data and has drawn comparative lines with the current TCM principles. The computational approach is also used to study molecular mechanisms like miRNA and siRNA interaction, glycolysis targeting, histone modifications, DNA methylation [7], [8], [21]. These mechanisms are particularly important to understand as they serve as the basis for the drug-target interaction studies. The present-day drugs are based on 'one drug fits all' theory which may not be a pharmacologically correct term for treating a disease. Network pharmacology is being used [13] for multicompound drug discovery. Diseases like cardiovascular disease, HIV, etc., that are being treated using TCM are being probed by computational methods into its mode of action [9].Various mathematical models have been suggested for network construction between herbs/natural products and their targets [12], [24], [15], [14], [16], [6]. The concept of network pharmacology is explained in detail in various review articles [16, 6].

The various steps involved in bioinformatics approaches are:

1. Text mining for collection of literature.
2. Data mining of the various available resources from databases. For example, 2D and 3D structure of compounds from chemspider, pubchem, etc.
3. Understanding the problem, finding loopholes, study design
4. QSAR studies
5. Target identification, target modeling
6. Network building, interpretation, visualization, pathway-enrichment analysis
7. Genome-wide association studies
8. Evolutionary analysis
9. Docking, molecular simulation

1.5 CONCLUSION

No science can justify independently its role in a biological system. Different areas of science have to come together for sharing scientific concepts and knowledge. Bioinformatics plays an important role in understanding biological mechanism and unveiling the underlying genomic entities responsible for causing diseases. Alternative and complementary medicine is presently gaining momentum. Ethnomedicine, phytochemistry, pharmacology of the plants, and its mechanism of action are being studied by using various molecular techniques and bioinformatics tools. The need of the hour is to fasten the research by using *in-silico* methods coupled with other biological techniques to scientifically validate the usage of the herbal medicines. Also there is a need for the researchers to gain the trust of the tribal people so as to utilize their knowledge. Study designs should be more comprehensive to fill the void in the medical knowledge.

KEYWORDS

- **Alternative medicine**
- **Bioinformatics**
- **Genomics**
- **Personalized medicine**

REFERENCES

1. Anonymous. Ayurveda and drug discovery. *Curr. Sci.* **2004**, *86*(6), 754.
2. Patwardhan, B.; and Khambholja, K.; Drug Discovery and Ayurveda: Win-Win Relationship Between Contemporary and Ancient Sciences, Drug Discovery and Development—Present and Future, Kapetanović, I., Ed.; InTech, **2011**; p 528.
3. Chatterjee, B.; and Pancholi, J.; *Prakriti*-based medicine: A step towards personalized medicine. *Ayu.* **2011**, *32*(2), 141–146.
4. Dunkel, M.; Fullbeck, M.; Neumann, S.; and Preissner, R.; SuperNatural: a searchable database of available natural compounds. *Nucl. Acid. Res.* **2006**, *34*, D678–D683.
5. Fang, Y. C.; Hsuan-Cheng Huang, H. C.; Chen, H. H.; and Juan, H. F.; TCMGeneDIT: a database for associated traditional Chinese medicine, gene and disease information using text mining. *BMC. Complem. Alter. Med.* **2008**, *8*, 58.
6. Fang, Z.; Zhang, M.; Yi, Z.; Wen, C.; Qian, M.; and Shi, T.; Replacements of rare herbs and simplifications of traditional Chinese medicine formulae based on attribute similarities and pathway enrichment analysis. Evidence-Based Complementary and Alternative Medicine, **2013**, 9 p.
7. Hsieh, H. Y.; Chiu, P. H.; and Wang, S. C.; Epigenetics in traditional Chinese pharmacy: A bioinformatic study at pharmacopoeia scale. *eCAM.* **2011**, 10 p.
8. Hsieh, H. Y.; Chiu, P. H.; Wang. S. C.; Histone modifications and traditional Chinese Medicinal. *BMC Comp. Alter. Med.* **2013**, 13, 115.
9. Hu, J. Z.; Bai, L.; Chen, D. G.; Xu, Q. T.; and William, M.; Southerland. Computational Investigation of the Anti-HIV Activity of Chinese Medicinal Formula Three-Huang Powder. *Interdiscip. Sci.* **2010**, *2*(2), 151–156.
10. Kang, G. L.; Shao, Li, S.; and Zhang, J. F.; Entropy-based model for interpreting life systems in traditional Chinese medicine. *eCAM.* **2008**, *5*(3), 273–279.
11. Kumar, D.; From evidence-based medicine to genomic medicine. *Gen. Med.* **2007**, *1*, 95–104.
12. Li, B.; Xu, X.; Wang, X.; Yu, H.; Li, X.; Tao, W.; Wang, Y.; and Yang, L.; A systems biology approach to understanding the Mechanisms of Action of Chinese Herbs for Treatment of Cardiovascular Disease. *Int. J. Mol. Sci.* **2012**, *13*, 13501–13520.
13. Li, J.; Lu, C.; Jiang, M.; Niu, X.; Guo, H.; Li, L.; Bian, Z.; Lin, N.; and Lu, A.; Traditional Chinese medicine-based network pharmacology could lead to new multicompound drug discovery. *eCAM.* **2012**, 11 p.
14. Li, S.; Network systems underlying traditional Chinese medicine Syndrome and Herb Formula. *Curr. Bioinform.* **2009**, *4*, 188–196.
15. Li, S.; Zhang, B.; Jiang, D.; Wei, Y.; and Zhang, N.; Herb network construction and co-module analysis for uncovering the combination rule of traditional Chinese herbal formulae. *BMC Bioinfor.* **2010**, *11*(Suppl 11), S6.
16. Li, S.; Zhang, B.; and Zhang, N.; Network target for screening synergistic drug combinations with application to traditional Chinese medicine. *BMC. Syst. Biol.* **2011**, *5*(1), S10.
17. López-Pérez, J. L.; Therón, R.; Olmo, D. E.; Díaz, D.; NAPROC-13: a database for the dereplication of natural product mixtures in bioassay-guided protocols. *Bioinformatics.* **2007**, *23*(23), 3256–3257.
18. Novick, P. A.; Ortiz, O. F.; Poelman, J.; Abdulhay, A. Y.; and Pande, V. S.; SWEETLEAD: An in silico database of approved drugs, regulated chemicals, and herbal isolates for computer-aided drug discovery. *PLoS ONE,* **2013**, *8*(11), e79568.
19. Song, Y. N.; Zhang, G. B.; Zhang, Y. Y.; and Su, S. B.; Clinical applications of Omics technologies on ZHENG differentiation research in traditional Chinese medicine. *eCAM.* **2013**, 11 p.

20. Vetrivel, U.; Subramanian, N.; and Pilla, K.; InPACdb—Indian plant anticancer compounds database. *Bioinformation.* **2009**, *4*(2), 71–74.
21. Wang, Z.; Wang, N.; Chen, J.; and Shen, J.; Emerging glycolysis targeting and drug discovery from ChineseMedicine in cancer therapy. *eCAM.* **2012**, 13 p.
22. Xue, R.; Fang, Z.; Zhang, M.; Yi, Z.; Wen, C.; and Shi, T.; TCMID: Traditional Chinese Medicine integrative database for herb molecular mechanism analysis. *Nucl. Acid. Res.* **2013**, *41*, D1089–D1095.
23. Ye, H.; Ye, L.; Kang, H.; Zhang, D.; Tao, L.; Tang, K.; Liu, X.; Zhu, R.; Liu, Q.; Chen, Y. Z.; Li, Y.; and Zhiwei Cao, Z.; HIT: linking herbal active ingredients to targets. *Nucl. Acid. Res.* **2011**, *39*, D1055–D1059.
24. Zhao, M.; Zhou, Q.; Ma, W.; and Wei, D. Q.; Exploring the Ligand-Protein networks in traditional Chinese medicine: Current databases, methods, and applications. *eCAM.* **2013**, 15 p.

CHAPTER 2

TREATING FUNGAL DERMATOPHYTIC INFECTIONS

M. SHARANYA[1] and R. SATHISHKUMAR[2]

[1]Department of Bioinformatics, Bharathiar University, Coimbatore - 641 046, India

[2]Department of Biotechnology, Salem Sowdeswari College, Salem - 636 010, India

CONTENTS

2.1 INTRODUCTION

Fungi are a group of eukaryotes which are unicellular or multicellular, or syncytial spore-producing organisms with approximate size ranging from 2 to 10 μm, and may be either beneficial or harmful. The fungal cell contains membrane bound organelles like nuclei, mitochondria, golgi apparatus, endoplasmic reticulum, lysosomes, etc., and remarkably ergosterols in the external membrane forming a rigid structure with the help of chitin molecules. Fungi possess 80S ribosomes and the cell division exhibit mitosis. Basic requirement of fungi includes water, and oxygen, and perhaps fungi are said as chemoheterotrophs (require organic compounds for both carbon and energy sources), osmotrophic (obtain nutrients by absorption), saprophytes (lives on decaying matter) or parasites (lives on living matter) and there are no obligate anaerobes. In general, lipids and glycogen are the storage form and the reproduction occurs either asexually and/or sexually.

Living organisms are classified into five kingdoms namely Monera, Protista, Fungi, Plantae and Animalia where the fungi are placed in a separate kingdom by R. H.Whittaker in 1969. Fungi are classified based on sexual/asexual reproduction and morphology. The fungi are grouped as zygomycetes (produce zygospore), ascomycetes (produce endogenous spores called ascospores in cells called asci), basidiomycetes (produce exogenous spores called basidiospores in cells called basidia) and deuteromycetes (a heterogeneous group of fungi where no sexual reproduction has yet been demonstrated and also called fungi imperfecti). Similarly on morphological basis, they are molds (e.g., *Aspergillus* spp., *Microsporum gypseum*, etc.), yeasts (e.g., *Cryptococcus neoformans, Saccharomyces cerviceae*), yeast-like (e.g., *Candida albicans*) and dimorphic (e.g., *Histoplasma capsulatum, Blastomyces dermatidis, Paracoccidiodes brasiliensis, Coccidioides immitis*).

2.2 FUNGAL INFECTIONS

Among an estimated 1.5 million species of fungi, some 200 have been recognized as "human pathogens" and the infections are normally categorized as follows:

- Superficial sicolor, black -(superficial phaeohyphomycosis, tinea versicolor, black piedra and white piedra)
- Cutaneous -(dermatophytosis and dermatomycosis)
- Subcutaneous mycetoma, sporotrichosis, -(chromoblastomycosis, rhinosporidiasis, subcutaneous phaeohyphomycosis, lobmycosis)

- Systemic domycosis,   -(blastomycosis, histoplasmosis, coccidioi-paracoccidioidomycosis)
- Opportunistic   -(candidiasis, cryptococcosis, aspergillosis)
- Other mycoses   -(otomycosis and occulomycosis)
- Fungal allergies   –(asthma and sinusitis)
- Mycetism and Mycotoxicosis   —(anorexia, edema of legs, massive gastro-intestinal bleeding)

The infections are mostly diagnosed by specimen collection and further followed by microscopy or using other techniques such as culturing, serology, antigen detection, skin tests and molecular techniques accordingly to the type and site of infection.

2.2.1 FUNGAL SKIN INFECTION

Various microscopic organisms live harmlessly inside the body and on the surface of the skin. However, certain types of fungus that are normally harmless, on overgrowth can cause superficial and systemic infections which are more commonly seen in those people undertaking antibiotics, corticosteroids, immunosuppressant drugs and contraceptives. Even also prevail in people with endocrine disorders, immune diseases, diabetes and others such as AIDS, tuberculosis, major burns and leukemia. Moreover found in obese people with excessive skin folds (Common Fungal Infections of the Skin, Spring 2006). In skin infections the topmost layer called stratum corneum is highly affected especially when exposed to a barrage of insults from the environment and thrive mostly on moist areas of the body, such as under the breasts, in the groin, and between fingers and toes. Some may cause no discomfort but others involve itching, swelling, and pain. Cutaneous fungal infections are often divided into "superficial" and "deep" forms [1]. The most common types of superficial infections are ringworm, athlete's foot and jock itch, and are approximately 25 percent of the populations afflicted with them (Common Fungal Infections of the Skin, Spring 2006). In the case of inflammatory conditions the fungi multiply and invade the skin, digestive tract, the genitals, or other body tissues. The most common fungal infections that infect the skin belong to a class of fungus called "tinea". Tinea refers exclusively to dermatophyte infections occur on skin, hair and nails (Continuum of Care, Judith Stevens). Noble and Forbes [4] quoted "among the most common skin diseases, specifically superficial fungal infections affect millons of people around the world".

2.2.2 DERMATOPHYTES

The dermatophytes are capable of invading keratinized tissues of humans and other animals to cause an infection, dermatophytosis, commonly referred to as ringworm but are not able to penetrate deeper tissues or organs of immunocompetent hosts [5]. Dermatophytes exist as both keratinophilic and keratinolytic and are often live on dead tissues. The infection may be mild to severe according to the host immune condition and reaction to the metabolic products of the fungus, the virulence of the infecting strain or species, the anatomic location of the infection and the local environmental factors. Dermatophytes as saprophytes reproduce asexually *via* sporulation of anthro-, micro- and macroconidia produced from the specialized conidigenous cells. Vegetative structures are observed with typical arrangement on hyphae, chlamydospores, spirals, antlershaped hyphae (chandeliers), nodular organs, pectinate organs and racquet hyphae. Growth forms and pigmentation produced by the dermatophyte colonies would be the presumptive identification of the species where five most important colony characteristics had been listed by Ajello [6], they are (1) rate of growth (2) general topography (flat, heaped, regularly or irregularly folded), (3) texture (yeast-like, glabrous, powdery, granular, velvety or cottony) (4) surface pigmentation and (5) reverse pigmentation.

According to WHO (World Health Organisation), the dermatophytes are defined in three genera: *Epidermophyton, Trichophyton* and *Microsporum* [7]. They comprise about 40 different species, and have common characteristics:
1. Close taxonomic relationships
2. Keratinolytic properties (they all have the ability to invade and digest the keratin as saprophytes "*in vitro*" and as parasites "*in vivo*", producing lesions in the living host).
3. Occurrence as etiologic agents of infectious diseases of man and/or animals

Ecologically dermatophytes are grouped as
i)   zoophilic   -   Keratin utilizing on hosts- found in living ani
                             mals (e.g., *M.canis, T.verrucosum*);
ii) anthropophilic   -   Keratin utilizing on hosts- found in humans
                             (e.g., *M.audounii, T.tonsurans*);
iii) geophilic   -   Keratin utilizing soil saprophytes (e.g.,
                             M.gypseum, T.ajelloi)

The dermatophytic infections are named according to the site of infection (Table 2.1).

TABLE 2.1 Nosology of dermatophyte infections

S. No.	Infection Name	Area of Infection	Organism	Description
1.	Tinea capitis	Scalp infection	*M.canis, M.audounii, T.tonsurans, T.verrucosum*	Small papule spreads to form scaly, irregular or well-demarcated areas of alopecia. The cervical and occipital lymph nodes may be enlarged; a kerion, a boggy, inflammatory mass, followed by healing.
2.	Tinea corporis	Trunk and limb infection	*T.rubrum, T.verrucosum, M.canis*	Single or multiple scaly annular lesions with a slightly elevated, scaly and or erythematous edge, sharp margin and central clearing.
3.	Tinea barbae	Hair infection	Zoophilic, *Trichophyton* sp.	Patches of inflammation, sometimes with follicular pustules in beard area
4.	Tinea faciei	Facial infection	Zoophilic, *Trichophyton* sp.	Patches of inflammation, sometimes with follicular pustules in surface of the face
5.	Tinea cruris	Groin infection	*T.rubrum, T.interdigitale, E.floccosum*	Most common in men, typical lesion is red, marginated eruption which spreads outwards from groin crease with edges may scaly, pustular or vesicular
6.	Tinea pedis	Foot infection	*T.rubrum, T.interdigitale, E.floccosum*	Occur in three distinct forms; interdigital toe webs fissured, macerated and itchy, vesicular patches affect soles and sides of feet lead to blisters and itchy, dry scaly changes over whole plantar surface and extending up the sides of the feet, producing a demarcated line (moccasin pattern)
7.	Tinea manuum	Hand infection	*T.rubrum* (rarely geophilic may also occur)	Unilateral, diffuse scaling of the palm.
8.	Tinea unguium	Nail infection	*T.rubrum, T.interdigitale*	Mostly toe nail infected, nails separate from the nail-bed, the nail plate thickens and crumbly and yellow-brown.

2.2.3   CURRENT TREATMENTS

The infection by dermatophytes shows acute and chronic inflammatory changes in the dermis and can be treated with the antifungal agents either by topical application or by oral intake. Topical agents should posses the ability to penetrate the stratum corneum cells whereas the oral treatment would be suitable in the case of inflammatory infections and hyperkeratotic lesions [7].

The existing classes of antifungal agents are polyenes, azoles and pyrimidines, others include allylamines, candins and the drug griseofulvin each targets the fungal cell in their unique mode of action (see Figure 2.1). Flucytosine, belonging to the pyrimidine class of antifungal agent, is the only drug targeting the thymidylate synthetase and thereby interrupt in the DNA synthesis [8]. On long usage, flucytosine exhibit toxicity to bone marrow which further leads to anemia, leucopenia, thrombocytopenia and also causes nausea, vomiting, diarrhea, liver damage, nephrotoxicity, abdominal cramps and pain. Allylamines and azoles inhibit the ergosterol biosynthesis pathway where allylamines specifically reduces the squalene oxide formation and the azoles on 14-α-demethylase thereby facilitating the accumulation of lanosterol [9]. Use of azoles and allylamines for long time may cause hepatitis, teratogenic, allergic rash, hormone imbalance, nausea, vomiting and fluid retention. Polyenes disrupt the fungal cell membrane and pave the way for cell content leakage. However, prolong usage would result in infusion related events such as fever, chills, headache, nausea, vomiting and dose-limiting nephrotoxicity [10]. Echinocandin, is the only available semisynthetic class of drug in the market is also known as fungal (1,3)-β-D-glucan synthase inhibitors which associate in disruption of cell wall biosynthesis [11].

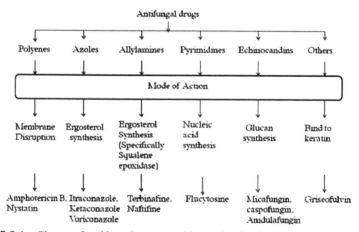

FIGURE 2.1   Classes of antifungal agents and its mode of action.

2.3   MEDICINAL PLANTS AS ALTERNATIVE AND COMPLEMENTARY MEDICINE

Medicinal plants play a prominent role in treating most number of diseases. The World Health Organization (WHO) estimates that over 80 percent of people in developing countries depend upon traditional medicine for treatment of diseases and woe in their primary health care [8812]. Since the existing drugs are determined with side effects and are also be reasoned for reoccurrence of the infection either at same or at different site of the body and in the development of fungal resistance. Therefore, alternative and complementary medication to treat dermatophytosis is mandatory. India, is a rich source of medicinal plants in which Western Ghats is considered as one of the hotspot of the country. At present, about 40 percent or 60,000 sq km of the Western Ghats is declared as an Ecologically Sensitive Area (ESA). According to the survey conducted by Krishnan et al. [13] one third of the plant species are being endemic and about 500 species are categorized with medicinal importance and has put forth several measures in the flora and fauna conservation. A project conducted in the period of 2005–2008 had surveyed and recorded the total number of medicinal plants along with their botanical details from the area of Western Ghat, has been considered and the number of plants reported scientifically for its antidermatophytic activity are accounted for its possible usage as an alternate and complementary medicine (Project by Kholkute, 2005–2008, submitted to ICMR).

2.3.1   WESTERN GHATS

The biogeographic zone of Western Ghats includes a narrow stretch along India's west coast approximately 30–50 km inland, starting from the hills south of the Tapi river in the north to Kanyakumari in the south. Western Ghats is otherwise called as "Sahyadri" traverse the States of Kerala, Tamil Nadu, Karnataka, Goa, Maharashtra and Gujarat (see Figure 2.2). These mountains cover an area of around 140,000 km² in a 1,600 km long stretch that is interrupted only by the 30 km Palghat Gap at around 11°N. The Ghats rise up abruptly in the west to a highly dissected plateau up to 2,900 m in height, and descend to dry Deccan plains below 500 m in the east. The extreme climatic and altitudinal gradient has resulted in a variety of forest types, from evergreen to semi-evergreen, from moist deciduous to dry deciduous formations. According to some biogeographers [14] the Western Ghats forms the "Malabar" province and are internationally recognized as a region of immense global importance for its outstanding features and enormous biodiversity of ancient lineage. Although the Western Ghats cover only 5 percent of India's total geographical area, the region contains over 30 percent of the country's plant species of which around 12,000 species, from lower groups to flowering plants, are estimated to occur here.

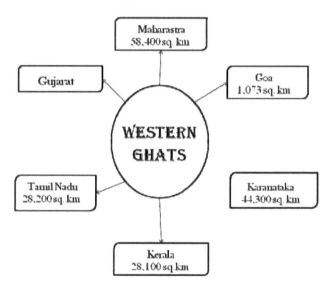

FIGURE 2.2   Area under Western Ghats.

2.3.1.1 MEDICINAL PLANTS REPORTED WITH ANTIDERMATOPHYTIC ACTIVITY

ACACIA MEARNSII

Acacia mearnsii De Wild. (Fabaceae) is found in abundance throughout Australia, Asia, Africa and America. The plant was known previously as *Racosperma mearnsii* and commonly known as Black Wattle tree, it is a short medium lived woody perennial and spreading tree, about 15 m high with smooth and greenish-brown bark on young branches which gradually turn blackish and rough on trunk. The young branchlets are downy. While it is widespread and common in lowlands, open forest, healthy woodland and on cleared land, particularly on dry, shallow soils [17], it grows in open forest, woodland or tussock grassland, in gullies or on hillsides, and in sandy or gravelly clay soils. It has a globular inflorescence with 20–30 tiny pale yellow flowers. Pods, dark brown to black in color, are more or less straight, 5–10 cm long, 5–8mm wide and strongly constricted between seeds [16]. The seeds are reported for its antimicrobial activity where the minimal inhibitory concentration (MIC) and minimal fungicidal concentration (MFC) lies at 2,500μg/ml and 5,000μg/ml against both *T. mucoides* and *T. tonsurans* [15].

ACALYPHA INDICA

Acalypha indica L. belongs to Euphorbiaceae family, is an annual, erect herb, up to 1 m high. Leaves are 2.5–7.5 cm long, ovate or rhomboid-ovate, crenate-serrate. Flowers in numerous lax, erect, elongated axillary spikes, the male minute, clustered near the summit of the spike, the females scattered, surrounded by a large. Traditionaly the leaves mixed with common salt is applied to scabies and other skin diseases (Indian Herbal Remedies, Khare CP). The leaves extracted with water and ethanol showed activity against *T.rubrum* (water: MIC-9.3 μg/ml, MFC-9.3 μg/ml; ethanol: MIC-9.3 μg/ml, MFC-9.3 μg/ml), *T.mentagrophytes* (water:MIC-9.3 μg/ml, MFC-9.3 μg/ml; ethanol: MIC-9.3 μg/ml, MFC-9.3 μg/ml) with respective MIC and MFC values [19]. The study by Radhika et al. [43] also revealed the significant activity exhibited by the ethanol and ethyl acetate extract against *T.rubrum*, *T.mentagrophytes*, *M.gypseum* and *T.tonsurans* showing MIC and MFC at 250 μg/ml for all the organisms rather than the hexanic extract which exhibited at higher 1,000 μg/ml concentration.

ACHYRANTHES ASPERA

Achyranthes aspera L. (Amaranthaceae) is commonly known as Rough Chaff tree is an erect or procumbent, annual or perennial herb, 1–2m in height, often with a woody base, commonly found as a weed of waysides, on roadsides. Stems angular, ribbed, simple or branched from the base, often with tinged purple color, branches terete or absolutely quadrangular, striate, pubescent, leaves thick, 3.8–6.3 × 22.5–4.5 cm, ovate- elliptic or obovate-rounded, finely and softly pubescent on both sides, entire, petiolate, petiole 6–20 mm long, flowers greenish white, numerous in axillary or terminal spikes up to 75 cm long, seeds subcylindric, truncate at the apex, rounded at the base, reddish brown [12]. The plant is used in treating skin diseases especially its oil relieves all kind of skin problems and the plants is also called prickly flower (http://www.astrogle.com/). The method of extraction also had impact on the plants biological activity. Study by Londonkar et al. [21] depicts the scenario where the leaves extracted with chloroform, petroleum ether and methanol using infusion and maceration methods were evaluated for its antidermatophytic activity. Where the methanol extract obtained by infusion method has shown more significant inhibitory activity against the dermatophytes *T.rubrum* and

M.canis with 14 mm and 12 mm dia of zone of inhibition for *T.rubrum* and 12 mm dia for *M.canis*.

ACORUS CALAMUS

Acorus calamus L. (Araceae), which is commonly known as sweet flag, is a herbaceous perennial with a rhizome that is long indefinite branched, smooth, pinkish or pale green. Its leaf scars are brown white and spongy and it possess slight slender roots. The leaves are few and distichously alternate whose size was found to be between 0.7 and 1.7cm wide with average of 1cm. The flowers are 3–8cm long, cylindrical, greenish brown and contains multitude of rounded spikes covering it. The fruits are found to be small and berry like with few seeds [22]. Traditionally the plant is used as a promising immunomodulatory agent in the inflammatory skin diseases (Divya et al., 2011) besides novel terpenoid, 1–2, 4, 5 trimethoxy phenyl 1' methoxy propionaldehyde (TMPMP) isolated was scientifically tested and reported to cure tinea pedis in wistar rat model infected with *T.rubrum* [23].

AEGLE MARMELOS

Aegle marmelos L. Correa (Rutaceae) is commonly called Vilvam in Tamil, Bael, Bengal quince or stone apple and are often cultivated in temples for its leaves which are used in poojas. The leaves, stem, bark and fruits of this plant have long been used in traditional medicine. Bael is a slow growing, tough sub-tropical tree and is the only plant belonging to the genus Aegle [25]. The tree grows wild in well-drained soil and attains a size of about 12–15 m height even in the harsh and dry climates. The branches contain spines that are arrow and are upto an inch in size. The leaves are alternate borne singly or in twos or threes and are made up of three to five oval, pointed, shallowly toothed leaflets. The flowers are fragrant and are found in clusters along the young branches [26]. The plants are rich source of bioactive compounds, each exist with its unique activity. Besides, quantity of the compounds varies conceivably from plant to plant and even within the parts of the same plant. In such a way, leaves were collected, dried and subjected to cold extraction with water and 100 percent ethyl alcohol and the coarse powder was also successively extracted with various organic solvents like hexane, benzene, chloroform, ethylacetate, methanol and water. Different fractions were collected, filtered and evaporated to dryness and are further evaluated for its antifungal activity on dermatophytes such as *T.mentagrophytes*, *T.rubrum*, *M.canis*, *M.gypseum* and *E.floccosum*. MIC and MFC values observed for all the extractions were at 400 µg/ml concentration beneath the methanol fraction, ethanol and water extract exhibited at 200 µg/ml against *T.mentagrophytes*, *M.canis* and *E.floccosum* [66]. The susceptibility of every organism varies according to the type of organic solvents used for plant extraction. Therefore, attention has to be paid in remitting the crude plant extracts or plant compounds into the suitable medication.

ALOE VERA

Aloe vera L. Burm. f. belongs to the family Liliaceae, is a hardy perennial plant with turgid green leaves joined at the stem in a rosette pattern. Leaves are formed by a thick epidermis (skin) covered with cuticle surrounding the mesophyll. It has been used externally to treat various skin conditions such as cuts, burns and eczema. It is said to be a miracle plant because of its medicinal and traditional uses [28]. Mosunmola et al. [27] carried out antidermatophytic activity of *Aloe vera* juice using two methods such as agar disc and agar well diffusion and concluded that both the testing methods were reliable and exert reproducibility. The determined Minimum Inhibitory Dilution (MID) and Minimum Fungicidal Dilution (MFD) revealed the toxicity possessed by the juice obtained from different areas (Ilorin and Shagamu, Nigeria) against *E.floccosum*, *M.audouinii*, *T.mentagrophytes*, *T.rubrum*, *T.schoenleinii* and *M.ferrugineum*. The efficiency in curing dermatophytosis improved when associated with garlic pills in treating equines infected with *T.equinum* [29]. The study revealed the progressive improvement on every five-day observation and a desirable response was observed at the end of 25th day.

ALPINIA GALANGA

Alpinia galanga Willd. (Family- Zingiberaceae) is used in culinary, medication and cosmetics for centuries. It is commonly known as Rasna and Sugandhmula in Sanskrit and Arattai in Tamil. It is a perennial herb found commonly throughout the Western Ghats, Mysore, Goa, Malabar and Gujarat. Roots are adventitious, in groups, fibrous, persistent in dried rhizomes, about 0.5–2 cm long and 0.1–0.2cm in diameter and yellowish brown in color. Rhizomes are cylindrical, branched, 2–8 cm in diameter, longitudinally ridged with prominent rounded warts (remnants of roots) marked with fine annulations; scaly leaves arranged circularly, externally reddish brown, internally orange yellow; odor pleasant and aromatic; spicy and sweet in taste (Chudiwal et al., 2010). In 1985, Janssen and Scheffer found terpinen-4-ol is the most active compound from the essential oils of fresh and dried rhizomes of *A.galanga* and also identified the presence of acetoxychavicol acetate through mass spectroscopy (MS) and nuclear magnetic resonance (NMR). In addition, the compound showed significant inhibitory activity against tested dermatophytes with MIC value ranging from 50 to 250 μg/ml. The 95 percent ethanolic extract of *A.galanga* rhizome exhibited inhibitory action on the growth of *M.canis*, *M.gypseum*, *T.mentagrophytes* [31].

ALANGIUM SALVIFOLIUM

Alangium salvifolium (L.f) Wang. belongs to the family Alangiaceae and locally called as Ankolam [34]. The plant is distributed in dry regions, plains and lower hills in India, their roots are useful for external application in acute case of rheumatism, leprosy and inflammation and internal application in cases of bites of rabbit and dogs [35]. Water extract of the ground wood was evaluated for its antidermatophytic activity and it has been concluded that the plant can be included in the herbal preparation for the treatment of some dermatomycotic infections [33].

ANDROGRAPHIS PANICULATA

Andrographis paniculata Nees (Family-Acanthaceae) is commonly known as 'King of Bitters'. It is the most popular and extensively used in Ayurveda, Unani and Siddha medicines and also as home remedy for various diseases in Indian traditional as well as tribal medicine. It is an annual, profusely branched, erect herb, and 0.5–1.0 m in height with a tap root. Leaves are green, lanceolate, 3–7 cm × 1–2.3 cm in size, glabrous with slightly undulate margin, acuminate apex with a tapering base. Flowers are small and solitary; corolla whitish or light pink in color with hairs. Fruit, a capsule, linear, oblong and acute at both ends; seeds numerous [37]. The dermatophyte *E. floccosum* is more susceptible to ethanolic extract of *A. paniculata* leaves showing 74.6 percent of inhibition where as *T. rubrum* exhibited about 70.9 percent inhibition [36].

ANTHOCEPHALUS INDICUS

Anthocephalus indicus Rich. (Rubiaceae) is a large tree with a broad umbrella-shaped crown and straight cylindrical bole. The branches are characteristically arranged in tiers. The tree may reach a height of 45 m with a stem diameter of 100–160 cm and sometimes it has a small buttress up to 2 m high. The bark is gray, smooth and very light in young trees, but rough and longitudinally fissured in old trees. The branches spread horizontally and drop at the tip. The leaves are glossy green, opposite, simple sessile to petiolate, ovate to elliptical (15–50 cm long by 8–25 cm wide). The ethanolic and hot water extracts of the ripened fruit was found inhibiting the growth of *T.rubrum* at MIC value 2 mg/ml [38].

ARTEMISIA NILAGARICA

Artemisia nilagarica (Clarke) Pampan is commonly called as Indian worm wood, belongs to the family Asteraceae. It is an aromatic shrub, 1–2 m high, yellow or dark red small flowers, grows throughout India in hills up to 2,400 m elevation. It is erect, hairy, often half-woody and the stems are leafy and branched. The leaves are pinnately lobed, 5–14 cm long, gray beneath. Mugwort blossoms with reddish brown or yellow flowers. The flowers are freely small and stand in long narrow clusters at the top of the stem. The fruit (achene) is minute. The essential oils from the leaves was determined with significant antidermatophytic activity against *T.rubrum*, *M.canis* and *M.gypseum* at 500 µg/ml, 125 µg/ml and 62.50 µg/ml MIC values and the measured zone of inhibition were 26 mm, 28 mm and 29 mm respectively [39].

ASCLEPIAS CURASSAVICA

Asclepias curassavica L. (Asclepiadaceae) is an erect perennial shrub, growing over a meter high; stems are green at the base and reddish at the top. Stems and leaves exude a milky sap, when damaged. Leaves are opposite, dark green to reddish green, long and narrow, (6–15 cm), tapering to a point at both ends, and located toward the end of the stems. Small bunches of flowers grow at the ends of branches. They are red, with an orange center, the petals are curved backwards. The fruit is a long narrow pod, which splits open to set loose flat seeds with silky hairs at one end. Hexanic and methanolic extract of leaves and stems exhibited activity against *T.mentagrophytes* and *T.rubrum* at MIC value >8.0 and 2.0mg/ml concentration [41] respectively. Among the few medicinal plants evaluated, hexane, ethylacetate and methanol revealed the significant activity against all the organisms, *T.mentagrophytes*, *T.simii* and *E.floccosum* except *T.rubrum* ([40].

AZADIRACHTA INDICA

Azadirachta indica commonly known as neem (Meliaceae) is native of India and naturalized in most of tropical and subtropical countries are of great medicinal value and distributed widespread in the world. It is a tree 40–50 feet or higher, with a straight trunk and long spreading branches forming a broad round crown; it has rough dark brown bark with wide longitudinal fissures separated by flat ridges. The leaves are compound, imparipinnate, each comprising 5–15 leaflets. The compound leaves are themselves alternating with one another. It bears many flowered panicles, mostly in the leaf axils. The selel are ovate and about one cm long with sweet scented white oblanciolate petals. It produces yellow drupes that are ellipsoid and glabrous, 12–20 mm long. Fruits are green, turning yellow on ripening, aromatic with garlic like odor. Fresh leaves and flowers come in March-April. Fruits mature between April and August depending upon locality [42].

Each and every parts of the neem triumph as a cure in almost all kind of diseases. Henceforth, the extraction of leaves and stem using water, ethanol and all kind of organic solvents exhibit significant toxicity against the pathogens [43] relative to the previous report showing least MIC value of about 0.57 μg/ ml by the ethanolic extract of stem ([19]. The maximum effect was observed in the neem seed and leaf against the pathogens *T.rubrum* and *Candida albicans* which are isolated from the HIV +ve infected immune deficiency patients [44].

BAUUHINIA VARIEGATE

Bauhinia variegate is commonly called orchid tree, is a semideciduous tree to 15 m (50 ft) tall, with a spreading crown. Leaves alternate, long petioled (to 3 cm [1.25 in] long), thin-leathery, simple but deeply cleft at apex, forming 2 large rounded lobes; lower surfaces downy, especially at top of petiole; blades with 11–13 veins extending from heart-shaped or rounded base. Flowers showy, fragrant, in few-flowered clusters near stem tips, appearing during leaf fall (early spring); 5 petals, clawed, overlapping, pale magenta to indigo (occasionally white), with dark red and yellow also on upper petal; 5 stamens (rarely 6). Fruit a flat, oblong pod, to 30 cm (1 ft) long, 10–15-seeded. *T.mentagrophytes* and *T.rubrum* were found susceptible to the extracts obtained from the leaves showing activity between 40 and 55 percent [45].

BOERHAAVIA DIFFUSA

Boerhaavia diffusa (Nictaginaceae) is a perennial creeping weed, prostrate or ascending herb, up to 1 m long or more, having spreading branches. The roots are stout and fusiform with a woody root stock. The stem is prostrate, woody or succulent, cylindrical often purplish, hairy, and thickened at the nodes. Leaves are simple, thick, fleshy and hairy, arranged in unequal pairs, green and glabrous above and usually white underneath. Flowers are minute, subcapitate, and present in a group of 4–10 together in small bracteolate umbels, forming axillary and terminal panicles. These are hermaphrodite, pedicellate and white, pink or pinkish-red in color. Two or three stamens are present and are slightly exerted. The stigma is peltate. The achene fruit is detachable, ovate, oblong, pubescent, five-ribbed and glandular, anthocarpous, and viscid on the ribs ([48]. *B.diffusa* root extracts extracted with different solvents exerts varying degree of inhibitory action on the *Microsporum* species [47, 67].

CASSIA TORA

Cassia tora Linn. (Family- Caesalpiniaceae) is distributed throughout India, Sri Lanka, West China and tropics. It is an annual herbaceous fetid herb, almost an undershrub, up to 30-90 cm high, with pinnate leaves. Leaflets are in three pairs, ipposite, obovate, oblong with oblique base and up to 10 cm long. Flowers are in pari in axils of leaves with five petals and pale yellow in color. Pods are somewhat flattened or four angled, 10–15 cm long and sickle shaped, hence commonly named as sickle-pod. The seeds are 30–50 in a pod, rhombohedral and gathered in autumn [49]. The leaves were extracted with cold methanol and observed for the antifungal activity where it showed strong inhibition for the growth of *M.canis* with 10 mm zone of inhibition and also inhibited the *T.rubrum* and *T.mentagraphytes* [1]. The organic extract of leaves specifically extracted with petroleum ether showed activity against *T.mentagrophytes* and *E.floccosum* showing about 10 and 20 mm of zone of inhibition [51]. Therefore the extract has to be analyzed for its bioactive compound responsible for the activity.

COSTUS SPECIOSUS

Costus speciosus (Koen.) Smith belongs to the family zingiberaceae, is a perennial rhizomatous herb with erect or spreading stems. Leaves are simple, smooth, persistent, spirally arranged around the trunk. The leaves are sub sessile and appear dark green in color, elliptic or obovate in shape. The inflorescence is a spike around 10 cms long with large bracts in sub terminal position. Flowers are white in color, 5–6 cm long with a cup-shaped labellum and crest yellow stamens. Fruit is capsule and red in color. Seeds are black, five in number with a white fleshy aril [52]. Hexane extract of the plant and the isolated two compounds sesquiterpenoids showed good activity showing MIC values of 62.5 μg/ml, 31.25 μg/ml, and 125 μg/ml against *T.simii*, *T.rubrum* and *E.floccosum* respectively [53].

CRYPTOLEPIS BUCHANANI

Cryptolepis buchanani Roem & Schult, commonly known as jambupatra sariva in Sanskrit, is a large evergreen laticiferous, woody climbing, perennial shrub common especially in deciduous forest of sub-Himalayan tracts, Bihar, Orissa, and East Uttar Pradesh in Varanasi region (Sharma et al., 2012). The methanol and aqueous extracts were evaluated against the human dermatophytic fungi and demonstrated marked inhibitory activity against *T.rubrum* (Vinayaka et al., 2010).

JATROPHA CUCAS

Jatropha cucas L. or physic nut, is a bush or small tree (up to 5 m height) and belongs to the Euphorbia family. It has thick glorious branchlets, a straight trunk and gray or reddish bark, masked by large white patches. Leaves are of length and width of 6 to 15 cm with shallow lobes and are arranged alternately. Leaves were used traditionally as a medicine in treating dermatophytic infections. It was found with potential activity against the clinically collected dermatophytes- *Trichophyton*, *Microsporum* and *Epidermophyton* species at five different concentrations of 250, 200, 150, 100 and 50 mg/ml. The effective minimum inhibitory concentration was observed between 19.95 and 79.43 mg/ml by the ethanol extract [54].

LAWSONIA INERMIS

Lawsonia inermis L. is a much branched glabrous shrub or small tree (2 to 6 m in height). Leaves are small, opposite in arrangement along the branches, subsessile, about 1.5–5 cm long, 0.5 to 2 cm wide, greenish brown to dull green, elliptic to broadly lanceolate with entire margin, petiole short and glabrous and acute or obtuse apex with tapering base. Young branches are green in color and quadrangular which turn red with age. Bark is grayish brown, unarmed when young but branches of older trees are spine tipped. Inflorescence is a large pyramid shaped cyme. Flowers are small, about 1 cm across, numerous, fragrant, white or rose colored with four crumbled petals. Calyx is with a 0.2 cm tube and 0.3 cm spread lobes. Fruit is a small brown colored round capsule. Fruit opens irregularly and splits into four sections at maturity and is many seeded. Seeds are about 3 mm across, numerous, smooth, pyramidal, hard and thick seed coat with brownish coloration. It is commonly called as Henna and belongs to the family Lythraceae (Chaudhary et al. 2012). The bark extract exhibited absolute toxicity against 13 ringworm fungi, the activity has remained the same even after autoclaving at high temperature and on long storage (57). MIC values ranging between 3.12 and 12.5 mg/ml demonstrates the natural antidermatophytic activity of the extracts [68] and on comparison to other plants such as *Juglans regia*, *Pistacia lentiscus*; *L.inermis* appeared more active in inhibiting the dermatophytes with 18.87 ± 0.58 mm of zone of inhibition [55].

MENTHA ARVENSIS

Mentha arvensis L. commonly called mint and belongs to the family Lamiaceae. The essential oil of *M.arvensis* when assessed against the *T.rubrum* and *M.gypseum* exhibited strong activity and the formulation as ointment combining with essential oils from *Chenopodium ambrosioides*, *Cymbopogon citrates*, *Caesulia axillaris* and *Artemisia nela-grica* were able to cure experimental ringworm in guinea pigs within 7–12 days (Kishore et al., 1993). Bringing two or more oils or crude extracts together, sometimes may forbid or may enhance the activity, the activity is named either as synergism or as antagonism. *M.arvensis* show evidence of possessing synergistic effect with the above mentioned essential oils.

MORINGA OLEIFERA

Moringa oleifera Lam. are eaten and is cultivated for foods and medicinal purposes (Olson, 2002). It is commonly called as horse radish, benzolive, drumstick. The plant is a perennial soft wood native to the sub-Himalayan tracts of India, Pakistan, Bangladesh, and Afghanistan. *Moringa oleifera* is a small, fast-growing evergreen or deciduous tree that usually grows up to 10 or 12 m in height. It has a spreading, open crown of drooping, fragile branches, feathery foliage of tripinnate leaves, and thick, corky, whitish bark. The essential oil of *M.oleifera* exhibited prominent antidermatophytic activity [59] where the GC-MS analysis was observed with the presence of 44 compounds. The leaf extracts were extracted with water, methanol and 70 percent ethanol, among them ethanolic extracts greatly minimized the growth of the tested dermatophytic organisms such as *M.ferrugineum*, *T.soudanensse*, *T.tonsurans*, *T.verrucosum* and *T.mentagrophytes* (60, Oluduro, 2012).

MURRAYA KOENIGII

Murraya koenigii Spreng. belongs to the family Rutaceae, commonly known as curry-leaf tree, is a native of India, Sri Lanka and other south Asian countries. It is found almost everywhere in the Indian subcontinent, it shares aromatic nature, more or less deciduous shrub or tree up to 6 m in height and 15–40 cm in diameter with short trunk, thin smooth gray or brown bark and dense shady crown [62]. The ethanolic extract exerted significant effect on the hyphal morphology, conidiation and germination of *T.mentagrophytes* and *M.gypseum*. The assessed total lipid and ergosterol content were found to have decremented compared to the normal level, hence could be believed that the crude extract had influence on both the lipid and sterol synthesis. It was also found to inhibit the lipase secretion in the tested organisms [61]. From ancient time, the curry leaves are employed as an ingredient in day-to-day food and these reports have proven the medicinal value of Indian foods.

OCCIMUM GRATISSIMUM

Occimum gratissimum L. is an aromatic, perennial shrub belonging to the family Lamiaceae. It is commonly known as Scent leaf or Clove basil and is found in many tropical countries. It is 1–3 m tall; stem erect, round-quadrangular, much branched, glabrous or pubescent, woody at the base, often with epidermis peeling in strips. Silva et al. [64] evaluated and reported the antifungal activity of the hexanic fraction and the pure compound eugenol against the *M.canis*, *M.gypseum*, *T.rubrum* and *T.mentagrophytes*. Growth was completely inhibited by the hexane extract at the concentration of 125 µg/ml, whereas the eugenol shows only 80% of inhibition. The potentiality of the leaves on the dermatophytic inhibition was also stated by Mbakwem-Aniebo et al. [63].

OCCIMUM SANCTUM

Occimum sanctum L. is commonly called Basil an annual herb belonging to the mint family has been cultivated for thousands of years and has become an essential ingredient in many cooking traditions. Basil is a member of the Lamiaceae, used both as a culinary and ornamental herb. The organic extracts of the plant leaves exhibited broad spectrum of inhibitory action against most of the dermatophyte species such as *T.rubrum*, *T.mentagrophytes*, *T.tonsurans*, *M.canis*, *E.floccosum*, *M.nanum* and *M.gypseum* [65]. MIC and MFC were calculated based on the NCCLS method and the particular fraction obtained from methanol extract showed precise activity against *T.mentagrophytes* with MIC value about 125 ± 25 µg/ml concentration [24].

PHYLLANTHUS AMARUS

Phyllanthus amarus Schum & Th. (Euphorbiaceae) is an annual, glabrous herb grows up to 15–60 cm high. Has an errect stem, naked below and slender and spreading leaf branches above. Leaves are numerous, subsessile, pale green, often distichously imbricating, glaucous below, elliptic to oblong, obtuse, and stipules subulate. Flowers arise in leaf axis, very numerous, males 1–3 and females solitary. Sepels of male orbicular and obovate to oblong in females. Stamens 3, anthers sessile and in a short column. Disc of male minute glands and of females annular and lobed. Capsules depressed globose, smooth and hardly 3 lobed. Seeds are 3-gonous, rounded and with longitudinal regular parallel ribes on the back. The chloroform extract of aerial parts showed maximum inhibition against the dermatophyte *M.gypseum* at 4,000ppm concentration and greatly reduced the sporulation of the organism [46, 47].

PIPER BETLE

Piper betle L. belongs to the family Piperaceae, is commonly used as a cultural symbolism and the leaves of this plant are economically and medicnally important. Water extracts showed MIC of 9.3mg/ml against *T.rubrum* and *T.mentagrophytes* [19]. Since the ethanolic extract of *P.betle* leaves showed promising activity against the zoonotic dermatophytes (*M.canis*, *M.gypseum* and *T.mentagrophyte*), Trakranrungsie et al. [69] has formulated the extract into cream (Pb cream). The Pb cream containing 80µg of *P.betle* extract revealed comparable zones of inhibition with the ketoconazole. It was found that the bioactive compound hydroxychavicol from the chloroform fraction collected from aqueous extract exhibited antidermatophytic activity showing MIC value between 7.81 and 62.5µg/ml [70]. The chloroform extract showed 46 mm zone of inhibition against the *T.tonsurans* whereas the organism become more susceptible when treated with extract containing both chloroform extract of *P.betle* and *Allamanda cathertica* have increased the zone to 51 mm [68].

PIPER LONGUM

Piper longum L. (Family: Piperaceae) grows all over India, in evergreen forests and is cultivated in Assam, Tamil Nadu and Andhra Pradesh. A small shrub with a large woody root and numerous creeping, jointed stems, thickened at the nodes. The leaves are alternate, spreading, without stipules and blade varying greatly in size. The lowest leaves are 5–7 cm long, whereas, the uppermost 2–3 cm long. The flowers are in solitary spikes. The fruits, berries, in fleshy spikes 2.5–3.5 cm long and 5 mm thick, oblong, blunt and blackish green in color. The mature spikes collected and dried, form the commercial form of pippali and the root radix is known as pippalimula. The chloroform extract of the leaf showed better activity when compared to the petroleum ether, methanol and water extracts against *T.rubrum*, *T.tonsurans*, *M.fulvum* and *M.gypseum* where the MIC was recorded at 5mg/ml. The major compound showing the bioactivity was identified as 1,2-benzenedicarboxylic acid bis-(2-ethylhexyl) ester, 2,2-di-methoxybutane, and β-myrcene obtained through the analysis of GC-MS data from the fractions collected using silica gel column chromatography [71].

POGOSTEMON PARVIFLORUS

Pogostemon parviflorus Benth. belongs to the family Lamiaceae, is a suffruti-cose shrubm 1.2–1.8m high, stem and branches obtusely quadrangular, usually purple. Leaves 7.5–18 cm long, broadly ovate, acute or acuminate, coarsely and irregularly doubly-toothed, base cuneate. Flowers in dense pubescent spikes, forming pyramidal lax panicles. Corolla white, stamens exerted, filaments pur-ple except just below the anthers, bearded with purple hairs. The ethanolic ex-tract of *P.parviflorus* leaf completely prohibited the growth of *T.mentagrophytes*, *M.canis* and *M.gypseum* with minimum inhibitory concentration (MIC) values between 2.5 and 10 mg/ml [72].

PSIDIUM GUAJAVA

Psidium guajava is a medium sized tree with evergreen, opposite, aromatic short-petioled leaves. The inflorescence axillay 1–3 flowered trees are used for treatment of various disease conditions especially in the developing countries (74). Hexane extracts of *P.guajava* leaves was found to inhibit *Trichophyton ru-brum*, *Trichophyton tonsurans*, *Sporotrix schenckii*, *Microsporum canis* show-ing zone of inhibitions at or greater than 10mm [73].

PUNICA GRANATUM

Punica granatum L. (Punicaceae) is a small multistemmed shrub/tree, 5–10 m tall, canopy open, crown base low, stem woody and spiny, bark smooth and dark gray. Leaves are simple, 2–8 cm long, oblong or obovate, glabrous, oppositely placed, short-petioled shining.surface. Flowers regular, solitary, or in fascicles at apices, 4–6 cm. Petals lanceolate, 5–7 cms, wrinkled, and brilliant orange-red. Hypanthium colored, 58 lobed. Anthers numerous. Calyx persistent. Fruit a round berry, 5–12 cm, pericarp leathery. Interior compartmentalized with many pink-red sections of pulp-like tissue, each contains a seed grain. Fruits globose with persistent callipe and a coriaceous woody rind. Seeds numerous, angular with fleshy testa, 1.3 cm long [77]. In 2008, Dutta et al. found that water ex-tract of *P.granatum* was detrimental to dermatophytes and this was confirmed

with result obtained by Shrivastav et al. [76] against the growth on *T.tonsurans,*
T.mentagrophytes, T.rubrum, T.equinum, M.gypseum, M.nanum, M.audouinii.

ROSMARINUS OFFICINALIS

Rosmarinus officinalis L. belongs to the family Lamiaceae, is a small evergreen
which grows wild in most Mediterranean countries, reaching a height of 1.5m.
The main producers are Italy, Dalmatia, Spain, Greece, Turkey, Egypt, France,
Protugal and North Africa. Essential oils of R.officinalis, known as rosemary
oils, are obtained by steam distillation of the fresh leaves and twigs, and the
yields range from 0.5 to 1.0 percent. Color of the oil is almost colorless to yellow liquid with a pleasant odor and the major constituents are described as
α-pinene, 1,8-cineole and camphor (Tiwari and Virmani, 1987). Rosemary oil
exhibited strong antidermatophytic activity against the *M.audounii*, *T.rubrum*,
T.violaceum, *T.tonsurans*, *T.verrucosum*, *T.mentagrophyte*, and *E.floccosum* at
1 percent concentration [79].

RUBIA CORDIFOLIA

Rubia cordifolia L. is a climbing or scrambling herb, with red rhizomatous base
and roots. The plant is commonly known as 'Indian Madder' and sold under the
trade name 'manjistha'. Stem is quadrangular, divaricately branched, glabrous
or prickly-hispid, especially on the angles. Leaves are 3.8–9 × 1.6–3.5 cm long,
arranged in a whorl of four, cordate-ovate to ovate-lanceolate, 3–9 palmately
veined, upper surface mostly glabrous and rough [81]. The petroleum ether
extract of root exhibited antifungal activity against *T.rubrum* showing MIC at
0.253 mg/ml and further the extract was fractionized and analyzed through TLC
which showed the presence of anthraquinones [80].

SOLANUM INDICUM

Solanum indicum Linn. (Synonym: Solanum anguivi) belongs to the family Solanaceae commonly known as Byakur, is a bushy herb containing prickly spikes
in the stem and available throughout the India and all over the tropical and sub-
tropical regions of the world [83]. The leaves extracted with chloroform, methanol and water were tested against five dermatophytes such as *T.mentagrophytes*,
T.rubrum, *T.tonsurans*, *M.gypseum*, and *M.fulvum* and found high activity in the
chloroform extract showing MIC value about 2.5 to 5 mg/ml, followed by the
methanol extract, whereas the water extract exhibit in converse [82].

SOLANUM NIGRUM

Solanum nigrum L. (Solanaceae) is commonly called black nightshade, is an annual herbaceous plant which can reach up to 100 cm in height. The stem may be smooth or bear small hairs. The flowers usually white in color, have five regular parts and are up to 0.8 cm wide. The leaves are alternate and some-what ovate with irregularly toothed wavy margin and can reach 10cm in length and 5cm in width. The fruit is a round fleshy berry up to 2 cm in diameter and yellowish when ripe. The seeds are brown and numerous [85]. Ali-shtayeh and Abu Ghdeib, [84] reported the antidermatophytic activity against *M.canis*, *T.mentagrophytes* and *T.violaceum* showing MIC value about 8.0 ± 2.82, 61.5 ± 10.13 and 81.2 ± 10.83µg/ml respectively. Ethanolic and aqueous extract sig-nificantly inhibited the *Trichophyton* species and no effect on *Epidermophyton floccosum* [86].

TAMARINDUS INDICA

Tamarindus indica L. of the Fabaceae, is an important food in the tropics. It is a frost-tender, tropical, evergreen tree, commonly called tamarind. The tree is densely foliated with pale green, compound, feathery leaflets which give the broad, spreading crown a light, airy effect. Tamarind may reach heights of 65 feet and a spread of 50 feet but is more often seen smaller. The delicate leaflets cast a diffuse, dappled shade which will allow enough sunlight to penetrate for a lawn to thrive beneath this upright, dome-shaped tree [88]. *Eucalyptus glob-ulus* and *Tamarindus indica* both produce allelochemicals which are already known for its antidermatophytic activity against *M.gypseum*, *T.terrestre*, and *M.gypseum* [87].

TERMINALIA ARJUNA

Terminalia arjuna belongs to the family Combretaceae and is a tree with simple leaf, smooth and thick bark. Flowers are small, regular, sessile, cup-shaped, polygamous, white, creamy or greenish-white, and robustly honey-scented. The inflorescence are short axillary spikes or small terminal panicles and fruits are obovoid-oblong, dark brown to reddish brown fibrous woody, indehiscent drupe [90]. The bark of the plant extracted with acetone, 95% alcohol and methanol were evaluated against five dermatophyes such as *T.mentagorphytes*, *T.rubrum*, *T.tonsurans*, *M.gypseum*, and *M.fulvum* using agar cup diffusion method. The methanol extract exhibited significant inhibition and showed MIC value about

75 µg/ml for *T.tonsurans*, 125 µg/ml for *T.mentagrophytes*, 250 µg/ml for *T.rubrum*, *M.fulvum*, and 2,000 µg/ml for *M.gypesum* [89].

THYMUS VULGARIS

Thymus vulgaris L. (family: Labiatae or Lamiaceae) also known as common thyme, a plant native to the Mediterranean region has long been used as a source of the essential oil (thyme oil) and other constituents (e.g., thymol, flavanoid, caffeic acid, and labiatic acid) derived from the different parts of the plant [92]. The essential oil of *T.vulgaris* contains thymol as a major component and it was observed that the oil with antidermatophytic activity against *T.mentagrophytes* [91].

TRIDAX PROCUMBENS

Tridax procumbens Linn belongs to the family Compositae. It is commonly known as 'Common button' or 'Coat button' and it is a weed found throughout India. A hispid, procumbent herb with woody base sometimes rooting at the node, up to 60 cm high. Leaves are ovate-lanceolate 2–7 m and lamina pinnatisect, sometimes three lobed; flowers in small, long peduncled heads; achenes 1.5–2.5 mm long x 0.5–1 mm in diameter and densely ascending pubescent; persistent ; bristles of disc achenes alternately longer and shorter; and 3.5–6 mm in length [94]. *T.procumbens* along with *Lantana camara* and *Capparis decidua* in the form of ointment were applied topically in the *T.mentagrophytes* infected animal model. The *in vitro* analysis showed the root and leaf extracts were effective against the pathogen showing MIC value at 0.312mg/ml and 0.625mg/ml [93].

WRIGHTIA TINCTORIA

Wrightia tinctoria R. Br. belongs to the family Apocynaceae and is commonly called "indrajav." It is a small and deciduous tree which grows up to 10m with milky latex, scaly, smooth, and ivory-colored bark. Leaves are about 8–15cm, opposite, variable, elliptic lanceolate, or oblong lanceolate. Leaves are acute or rounded at the base, acuminate at the apex, petioles 5 mm long. Flowers are usually seen at the tip of branches with 6cm long cymes, white with fragrance. It is widely distributed in India and Burma [98]. It was found that the leaves had potent antidermatophytic activity against most of the organisms such as *T.tonsurans*, *T.mentagrophytes* in addition to *T.rubrum* and *E.floccosum* showing IC_{50} value at 2mg/ml [95].

Ponnusamy et al. [96] tested the hexane and chloroform extracts of leaves against *T.rubrum*, *E.floccosum* at 0.5 mg/ml concentration and the major compound indirubin also found to exhibit activity against *E.floccosum* (MIC- 6.25 µg/ml), *T.rubrum*, and *T.tonsurans* (MIC- 25µg/ml). Leaves also extracted with ethanol completely inhibit the pathogen *E.floccosum* at 500–100 ppm [97].

ZINGIBER OFFICINALE

Zingiber officinale Roscoe (Zingiberaceae), commonly called ginger, is a perennial rhizome which creeps and increases in size underground. Ginger is highly medicinal and used in the food preparations. Ginger is native to China and India. The essential oil of *Z.officinale* was found with strong antifungal activity against *T.rubrum* and *M.gypseum* showing the zone of inhibition about 72 mm and 69 mm and MIC value at 0.05 µg/ml and 0.06 µg/ml, respectively. Whereas, the ginger oil in mixture with turmeric oil exhibited an excellent activity (82mm, MIC 0.02 µg/ml against *T. rubrum* and 79 mm, MIC 0.04 µg/ml against *M. gypseum*), emphasizing its synergistic potentiality (Meenakshi and Richa, 2010).

2.4   CONCLUSION

Herbal remedies are considered the oldest forms of health care known to mankind on this earth. In India, it is reported that traditional healers use 2,500 plant species for curing several types of ailments and about 100 species of plants regularly serve as sources of medicine (Shenq-ji, 2001). Modern medicines basically depend on the traditional systems of medicine which is maintained and evolved over centuries among various tribal communities and are still maintained as a great traditional knowledge base in herbal medicines and retained by various indigenous groups around the world. Knowledge on the usage of herbal medicines is transferred from generation to generation. About 70 plant species belonging to 42 families are used for various purposes by tribal people of Western Ghats. This chapter is mainly focused on the flora of Western Ghats whose medicinal values were technically proven especially in treating the skin infection caused by a fungal group called Dermatophytes. In ICMR program a research group under the investigation of Dr. S.D. Kholkute surveyed and recorded the seasonal and nonseasonal plants along with the voucher specimens, botanical identity which resulted as 500 plants database. However, only 41 plants (i.e., 8.2% of the plants) have been found to be scientifically proven to be toxic against the dermatophytes. Therefore it is concluded that more effort is required in preserving our nature's gift and its appropriate utilization in healing various kind of ailments. Futuristic aspect is primarily to bring out the scientific knowledge on the synergism exhibited by plant extracts and oils, secondly to evaluate the Ayurvedic, Siddha, Unani, and other plant-based preparations for their bioactivity, thereby facilitating the improvement in the currently existing treatments.

KEYWORDS

- **Antidermatophytic activity**
- **Dermatophytes**
- **Fungal infections**
- **Herbal remedies**

REFERENCES

1. Kovacs, S. O.; and Hruze, L. I.; Superficial fungal infections. *Postgrad. Med.* **1995**, *98*(6), 61–75.
2. Anonymous. Common fungal infections of the skin. *Pharmawise*. Spring, **2006**, *10*(2), 1–6.
3. Judith Stevens, Fungal skin infections. Continuum of Care. Assessed date 25.7.2013.
4. Noble, S. L.; and Forbes, R. C.; Diagnosis and management of common Tinea infections. Kansas City, Missouri: American Academy of Family Physicians; **1998**.
5. Weitzman, I.; and Summerbell, R. C.; The dermatophytes clinic. *Microbiol. Rev.* **1995**, *8*(2), 240–259.
6. Ajello, L.; Georg, L. K.; Kaplan, W.; and Kaufman, L.; Laboratory manual for Medical Mycology; **1966**, US.
7. Palacio, A. D.; Garau, M.; Gonzalez-Escalada, A.; Calvo, M. T.; Trends in the treatment of dermatophytosis *Revista Iberoamericana de Micologia*; **2000**, 1(2),148–158.
8. Vermes, A.; Guchelaar, H. J.; and Dankert, J.; Flucytosine: a review of its pharmacology, clinical indications, pharmacokinetics, toxicity and drug interactions. *J. Antimicrob. Chemother.* **2000**, *46*, 171–179.
9. Ghannoum, M. A.; and Rice, L. B.; Antifungal agents: mode of action, mechanisms of resistance, and correlation of these mechanisms with bacterial resistance Clinic. *Microbiol. Rev.* **1999**, *12*(4), 501–517.
10. Gallis, H. A.; Drew, R. H.; and Pickard, W. W.; Amphotericin B: 30 years of clinical experience Ref. *Infect. Dis.* **1990**, *12*, 308–329.
11. Current, W. L.; Tang, J.; Boylan, C.; Watson, P.; Zeckner, D.; Turner, W.; Rodriguez, M.; Ma, D.; and Radding, J.; Glucan biosynthesis as a target for antifungals; the echinocandin class of antifungal agents. In The Discovery and Mode of Action of Adrugs; Dixon, G. K., Ed.; L. G Copping and D. W. Hollowmon, BIOS Scientific Publishers, Ltd.: Oxford, **1995**, pp 143–160.
12. Srivastav, S.; Pradeep Singh, Mishra, G.; Jha, K. K.; and Khosa, R. L.; *Achyranthes aspera*- An important medicinal plant: A review. *J. Nat. Prod. Plant. Resour.* **2011**, *1*(1), 1–14.
13. Krishnan, P. N.; Decruse, S. W.; and Radha, R. K.; Conservation of medicinal plants of Western Ghats, India and its sustainable utilization through *in vitro* technology. *In vitro Cell Dev. Biol. Plant,* **2011**, *47*, 110–122.

14. Roa, R. R.; and Sagar, K.; Invasive alien weeds of the Western Ghats: Taxonomy and Distribution–Chapter. 12. In Bhatt, J. R., Singh, J. S., Singh, S. P., Tripathi, R. S., Kohli, R. K. (Eds.) Invasive Alien Plants an Ecological Appraisal for the Indian Subcontinent; CABI Publishers, **2012**.

15. Olajuyigbe, Olufunmiso O.; Afolayan, Anthony, J. Pharmacological assessment of the medicinal potential of Acacia mearnsii De Wild.: Antimicrobial and toxicity activities. *Int. J. Mol. Sci.* **2012**, *13*, 4255–4267.

16. Maslin, B. R.; Introduction to Acacia. In Flora of Australia; Orchard, A. E., Wilson, A. J. G., Eds.; ABRS/CSIRO Publishing: Melbourne, Australia, volume 11A, **2001**; pp 3–13.

17. Walsh, N. G.; and Entwisle, T. J.; Flora of Victoria. Inkata Press: Melbourne, Australia, **1996**; Vol. 3.

18. Radhika, S. M.; and Michael, A.; Antidermatophytic activity of Azadirachta indica and Acalypha indica Leaves- An *in vitro* study. *Int. J. Pharm. Bio. Sci.* **2013**, *4*(4), 618–622.

19. Vaijayanthimala, J.; Rajendra Prasad, N.; Anandi, C.; and Pugalendi, K. V.; Anti-dermatophytic activity of some Indian med. *Plants.* **2004**, *4*, 1 26–31.

20. http://www.astrogle.com/ayurveda/curing-diseases-with-achyranthes-aspera-prickly-flower. html (assessed date- 20.11.2013

21. Londonkar, R.; Chinnappa Reddy, V.; and Abhay Kumar, K.; Potential antibacterial and antifungal activity of *Achyranthes aspera* L. *Recent Res. Sci. and Tech.* **2011**, *3*(4), 53–57.

22. Balakumbahan, R.; Rajamani, K.; and Kumanan, K.; *Acorus calamus*: An overview. *J. Med. Plants Res.* **2010**, *4*(25), 2740–2745.

23. Subha, T. S.; and Gnanamani, A.; Topical therapy of 1-2, 4, 5 trimethoxy phenyl 1' methoxypropionaldehyde in experimental *Tinea pedis* in Wistar rats. *Biol. Med.* **2011**, *3*(2), 81–85.

24. Balakumar, S.; Rajan, S.; Thirunalasundari, T.; and Jeeva, S.; Antifungal activity of *Aegle marmelos* (L.) Correa (Rutaceae) leaf extract on dermatophytes Asian Pac. *J. Trop. Biomed.* **2011**, *1*(4), 309–312.

25. Sharma, P. C.; Bhatia, V.; Bansal, N.; and Sharma, A.; A review on bael tree. *Nat. Prod. Rad.* **2007**, *6*, 171–178.

26. Maity, P.; Hansda, D.; Bandyopadhyay, U.; and Mishrra, D. K.; Biological activities of crude extracts and chemical constituents of bael, *Aegle marmelos* (L.). *Corr. Ind. J. Exp. Biol.* **2009**, *47*, 849–861.

27. Mosunmola, O. J.; RamotaRemi, R. A.; Abayomi, B. T.; Olaosebikan, M. S.; Charles, N.; Afolabi, O.; Susceptibility of dermatophytes to *Aloe vera* juices using agar diffusion and broth dilution techniques. *Am. J. Res. Comm.* **2013**, *1*(8), 53–62.

28. Rajeswari, R.; Umadevi, M.; Sharmila Rahale, C.; Pushpa, R.; Selvavenkadesh, S.; Sampath Kumar, K. P.; Bhowmik, D.; *Aloe vera*: The miracle plant its medicinal and traditional uses in India. *J. Pharmacog. Phytochem.* **2012**, *1*(4), 118–124.

29. Ferdowsi, H.; Afshar, S.; and Rezakhani, A.; A comparison between the routine treatment of equine dermatophytosis and treatment with Garlic-Aloe vera gel Int. Res. *J. Appli. Basic Sci.* **2012**, *3*(11), 2258–2261.

30. Chudiwal, A. K.; Jain D. P.; Somani R. S. *Alpinia galanga* Willd.—An overview on phytopharmacological properties Ind. *J. Nat. Prod. Res.* **2012**, *1*(2), 143–149.

31. Trakranrungsie, N.; Chatchawanchonteera, A.; and Khunkitti, W.; Ethnoveterinary study for antidermatophytic activity of *Piper betle, Alpinia galanga* and *Allium ascalonicum* extracts *in vitro. Res. in Vet. Sci.* **2008**, *84*, 80–84.

32. Janssen, A. M.; and Scheffer, J. J.; Acetoxychavicol acetate, an antifungal component of *Alpinia galangal. Planta. Med.* **1985**, *51*(6), 507–511.

33. Mansuang, W.; Sompop, P.; and Yuvadee, W.; Antifungal activity and local toxicity study of *Alangium salvifolium* subsp *hexaptalum*. Southeast. *Asian. J. Trop. Med. Pub. Health.* **2002**, *33*(3), 152–154.

34. Tariqo, S.; and Javed, A.; Vitamin C content of Indian medicinal plants- A review literature Ind. *Drugs*, **1985**, *23*(2), 72–75.
35. Bakhru, H. C.; Herbs that heal, Orient Longman Ltd., **1997,** p 17.
36. Ramya. R.; and Lakshmi Devi, N.; Antibacterial, antifungal and antioxidant activities of *Andrographis paniculata* Nees. Leaves. *Int. J. Pharmaceut. Sci. Res.* **2011**, *2*(8), 2091–2099.
37. Niranjan, A.; Tewari, S. K.; and Lehri, A.; Biological activities of Kalmegh (*Andrographis paniculata* Nees) and its active principles- A review. *Ind. J. Nat. Prod. Res.* **2010**, *1*(2), 125–135.
38. Mishra, R. P.; and Siddique, L.; Antifungal properties of *Anthocephalus cadamba* fruits. *Asian. J. Plant. Sci. Res.* **2011**, *1*(2), 81–87.
39. Vijayalakshmi, A.; Tripathi, R.; and Ravichandiran, V.;Characterization and evaluation of antidermatophytic activity of the essential oil from *Artemisia nilagirica* leaves growing wild in Nilgiris. *Int. J. Pharm. Pharm. Sci.* **2010**, *2*(4), 93–97.
40. Duraipandiyan, V.; and Ignacimuthu, S.; Antifungal activity of traditional medicinal plants from Tamil Nadu. *India. Asian. Pac. J. Trop. Biomed.* **2011**, S204–S215.
41. Garcia, V. M. N.; Gonzalez, A.; Fuentes, M.; Aviles, M.; Rios, M. Y.; Zepeda, G.; and Rojas, M. G.; Antifungal activities of nine traditional Mexican medicinal plants. *J. Ethnopharm.* **2003**, *87*, 85–88.
42. Hashmat, I.; Azad, H.; Ahmed, A.; Neem (*Azadirachta indica* A. Juss)—A nature's drugstore: An overview. *Int. Res. J. Biol. Sci.* **2012**, *1*(6), 76–79.
43. Radhika, S. M.; and Michael, A.; Antidermatophytic activity of *Azadirachta indica* and *Acalypha indica* Leaves- An *in vitro* study. *Int. J. Pharm. Bio. Sci.* **2013**, *4*(4), 618–622.
44. Santosh Kumar, S.; Anil, S.; Priyanka, S.; and Agrawal, R. D.; Antidermatophytic activities of different plant parts extract against *Trichophyton rubrum* and *Candida albicans* dermatophytes isolated from HIV+ Ves of Jaipur District Rajasthan Asian. *J. Biochem. Pharm. Res.* **2012**, *1*(2), 146–152.
45. Gunalan, G.; Saraswathy, A.; and Krishnamurthy, V.; Antimicrobial activity of medicinal plant *Bauhinia variegate* Linn. *Int. J. Pharm. Biol. Sci.* **2011**, 1(4), 400–408.
46. Agrawal, A.; Srivastava, S.; Srivastava, M. M.; Antifungal activity of *Boerhavia diffusa* against some dermatophytic species of Microsporum Hindustan Antibiot. *Bull.* **2004a**, (1–4), 45–46.
47. Agrawal, A.; Srivasta, S.; Srivastava, J. N.; and Srivastava, M. M.; Inhibitory effect of the plant *Boerhavia diffusa* L. against the dermatophytic fungus *Microsporum fulvum*. *J. Environ. Biol.* **2004b**, *25*(3), 307–311.
48. Sahu, A. N.; Damiki, L.; Nilanjan, G.; and Dubey, S.; Phytopharmacological review of *Boerhaavia diffusa* Linn. *(Punarnava) Phcog. Rev.* **2008**, *2*(4), 14–22.
49. Jain, S.; and Patil, U. K.; Phytochemical and pharmacological profile of *Cassia tora* Linn.— An Overview Ind. *J. Nat. Prod. Resour.* **2010**, *1*(4), 430–437.
50. Adamu, H. M.; Abayeh, O. J.; Ibok, N. U.; and Kafu, S. E.; Antifungal activity of extracts of some Cassia, Detarium and Ziziphus species against dermatophytes. *Nat. Prod. Radian.* **2006**, *5*(5), 357–360.
51. Rath, S.; and Mohanty, R. C.; Antifungal screening of *Curcuma longa* and *Cassia tora* on dermatophytes. *Int. J. Life Sci. Biotech. Pharm. Res.* **2013**, *2*(4), 88–94.
52. Rani, S.; Sulakshana, G.; and Patnaik, S. *Costus speciosus*, an antidiabetic plant. Review. *J. Pharm. Res.* **2012,** *1*(3), 52–53.
53. Duraipandiyan, V.; Al-Harbi, N. A.; Ignacimuthu, S.; Muthukumar, C.; Antimicrobial activity of sesquiterpene lactones isolated from traditional medicinal plant, *Costus speciosus* (Koen ex.Retz.) sm. *BMC Complement Alter. Med.* **2012**, *12*(13), 1–6.
54. Aniebo, C.; Okoyomo, E. P.; Ogugbue, C. J.; and Okonko, I. O.; Effects of *Jatropha curcas* leaves on common dermatophytes and causative agent of *Pityriasis versicolor* in rivers state, Nigeria Nature and Science. *2012*, 10(12), 151.

55. Mansour-Djaalab, H.; Kahlouche-Riachi, F.; Djerrou, Z.; Serakta-Delmi, S.; Hamimed, S.; Trifa, W.; Djaalab, I.; and Hamdipacha, Y.; *In vitro* evaluation of antifungal effects of *Lawsonia inermis*, *Psitacia lentiscus* and *Juglans regia*. *Int. J. Med. Arom. Plants*. **2012**, *2*(2), 263–268.

56. Sharma, K. K.; Saikia, R.; Kotoky, J.; Kalita, J. C.; Devi, R.; Antifungal activity of *Solanum melongena* L, *Lawsonia inermis* L. and *Justicia gendarussa* B. against dermatophytes. *Int. J. PharmTech. Res*. **2011**, *3*(3), 1635–1640.

57. Syamsudin, I.; and Winarno, H.; The effects of Inai (*Lawsonia inermis*) leave extract on blood sugar level: An Experimental Study *Res. J. Pharmacol*. **2008**, *2*(2), 20–23.

58. Chaudhary, G.; Goyal, S.; and Poonia, P.; *Lawsonia inermis* Linnaeus: A Phytopharmacological review. *Int. J. Pharm. Sci. Drug Res*. **2010**, *2*(2), 91–98.

59. Chuang, P. H.; Lee, C. W.; Chou, J. Y.; Murugan, M.; Shieh, B. J.; and Chen, H. M.; Antifungal activity of crude extracts and essential oil of *Moringa oleifera* Lam. *Bioresour. Technol.* **2007**, *98*, 232–236.

60. Ayanbimpe, G. M.; Ojo, T. K.; Afolabi, E.; Opara, F.; Orsaah, S.; and Ojerinde, O. S.; Evaluation of extracts of *Jatropha cucas* and *Moringa oleifera* in culture media for selective inhibition of saprophytic fungal contaminants. *J. Clinic. Lab. Anal*. **2009**, *23*, 161–164.

61. Jayaprakash, A.; and Ebenezer, P.; Antifungal activity of curry leaf (*Murraya koenigii*) extract and an imidazole fungicide on two dermatophyte taxa. *J. Acad. Indus. Res*. **2012**, *1*(3), 124–126.

62. Handral, H. K.; Pandith, A.; and Shruthi, S. D.; A review on *Murraya koenigii*: Multipotent medicinal plant. *Asian J. Pharm. Clinic. Res*. **2012,** *5*(4), 5–14.

63. Mbakwem-Aniebo, C.; Onianwa, O.; and Okonko, I. O.; Effects of *Ocimum gratissimum* leaves on common dermatophytes and causative agent of *Pityriasis versicolor* in rivers state, Nigeria. *J. Microbiol. Res*. **2012**, *2*(4), 108–113.

64. Silva, M. R.; Oliveira, J. G.; Femandes, O. F.; Passos, X. S.; Costa, C. R.; Souza, L. K.; Lemos, J. A.; and Paula, J. R.; Antifungal activity of *Ocimum gratissimum* towards dermatophytes. Mycoses. **2005**, *48*(3), 172–175.

65. Das, J.; Buragohain, B.; and Srivastava, R. B.; *In vitro* evaluation of *Ocimum sanctum* leaf extract against dermatophytes and opportunistic fungi Asian. *J. Microbiol. Biotech. Env. Sci.* **2010**, *12*(4), 789–792.

66. Balakumar, S.; Rajan, S.; Thirunalasundari, T.; and Jeeva, S.; Antifungal activity of *Ocimum sanctum* Linn. (Lamiaceae) on clinically isolated dermatophytic fungi Asian Pacific. *J. Trop. Med*. **2011**, 654–657.

67. Agrawal, A.; Srivastava, S.; Srivastava, J. N.; and Srivasava, M. M.; Evaluation of inhibitory effect of the plant *Phyllanthus amarus* against dermatophytic fungi *Microsporum gypseum* Biomed. *Environ. Sci*. **2004**, *17*(3), 359–365.

68. Sharma, K. K.; Saikia, R.; Kotoky, J.; Kalita, J. C.; Das, J.; Evaluation of antidermatophytic activity of *Piper betle*, *Allamanda cathertica* and their combination: an *in vitro* and *in vivo* study. *Int. J. Pharm. Tech. Res*. **2011**, *3*(2) 644–651.

69. Trakranrungsie, N.; Chatchawanchonteera, A.; and Khunkitti, W.; Antidermatophytic activity of *Piper betle* cream Thai. *J. Pharmacol*. **2006**, *28*(3), 16–20.

70. Ali, I.; Khan, F. G.; Suri, K. A.; Gupta, B. D.; Satti, N. K.; Dutt, P.; Afrin, F.; Qazi, G. N.; and Khan, I. A. *In vitro* antifungal activity of hydroxychavicol isolated from *Piper betle* L. *Annals of Clinic. Microbiol. Antimicrob*. **2010**, *9*(7), 1–9.

71. Das, J.; Jha, D. K.; Policegoudra, R. S.; Mazumder, A. H.; Das, M.; Chattopadhyay, P.; and Singh, L.; Isolation and characterization of antidermatophytic bioactive molecules from *Piper longum* L. leaves. *Ind. J. Microbiol*. **2012**, *52*(4), 624–629.

72. Sadeghi-Nejad, B.; and Deokule, S. S.; Antidermatophytic activity of *Pogostemma parviflorus* Benth. *Iran. J. Pharm. Res*. **2010**, *9*(3), 279–285.

73. Beatriz, P. M.; Ezequie, V. V.; Azucena, O. C.; and Pilar, C. R.; Antifungal activity of *Psidium guajava* organic extracts against dermatophytic fungi. *J. Med. Plants. Res.* **2012**, *6*(41), 5435–5438.

74. Geidam, Y. A.; Ambali, A. G.; and Onyeyili, P. A.; Preliminary phytochemical and antibacterial evaluation of crude aquesous extract of *Psidium guajava* leaf. *J. Appl. Sci.* **2007**, *7*, 511–514.

75. Dutta, B. K.; Rahman, I.; and Das, T. K.; Antifungal activity of Indian plant extracts: Antimyzetische Aktivitat indischer Pflanzenextrakte Mycoses. **2008**, *41*(11–12), 535–536.

76. Shrivastav, V. K.; Shukla, D.; Parashar, D.; and Shrivastav, A.; Dermatophytes and related keratinophilic fungi isolated from the soil in Gwalior region of India and in vitro evaluation of antifungal activity of the selected plant extracts against these fungi. *J. Med. Plants. Res.* **2013**, *7*(28), 2136–2139.

77. Arun, N.; and Singh, D. P.; *Punica granatum*: A review on pharmacological and therapeutic properties. *Int. J. of Pharm. Sci. Res.* **2012**, *3*(5), 1240–1245.

78. Tewari, R.; and Virmani, O. P.; Chemistry of Rosemary Oil: A Review. Central Institute of Medicinal and Aromatic Plants **1987**, *9*, 185–197.

79. Muyima, N. Y. O.; and Nkata, L.; Inhibition of the growth of dermatophyte fungi and yeast associated with dandruff and related scalp inflammatory conditions by the essential oils of *Artemisia afra, Pteronia incana, Lavandula officinalis* and *Rosmarinus officinalis. J. Essent. Oil-Bear. Plants.* **2005**, *8*(3), 224–232.

80. Gandhi, A. S.; Antifungal activity of *Rubia cordifolia* Pharmacology drug profiles **2006**, 1–6.

81. Devi Priya, M.; Siril, E. A.; Pharmacognostic studies on Indian Madder (*Rubia cordifolia* L.). *J. Pharmacognosy. Phytochem.* **2013**, *1*(5), 112–119.

82. Kotoky, J.; Sharma, K. K.; Kalita, J. C.; and Barthakur, R.; Antidermatophytic activity of *Solanum indicum* L. from North East India. *J. Pharm. Res.* **2012**, *5*(1), 265–267.

83. Chopra, R. N.; Nayer, S. L.; and Chopra, I. C.; Glossary of Indian Medicinal Plants. New Delhi: PID, **1992**, CSIR.

84. Ali-Shtayeh, M. S.; Abu Ghdeib, S. I.; Antifungal activity of plant extracts against dermatophytes Mycoses **1998**, *42*, 665–672.

85. Akubugwo, I. E.; Obasi, A. N.; and Ginika, S. C.; Nutritional potential of the leaves and seeds of Black Nightshade-*Solanum nigrum* L. *Var virginicum* from Afikpo-Nigeria Pakistan. *J. Nutr.* **2007**, *6*(4), 323–326.

86. Shamin, S.; Ahmed, S. W.; and Azhar, I.; Antifungal activity of Allium, Aloe and Solanum species. *Pharm. Biol.* **2004**, *42*(7), 491–498.

87. Sharma, R.; Upadhyaya, S.; Singh, B. S.; Singh, B. G.; Inhibitiory effect of allelochemicals produced by medicinal plants on dermatophytes. In Phytochemicals: a therapeutant for critical disease management; Khanna, D. R., Chopra, A. K., Prasad, G., Malik, D. S., Bhutiani, R. **2008**, pp 349–352.

88. Caluwe, E.; Halamova, K.; and Damme, P.V. *Tamarindus indica* L.- A review of traditional uses, phytochemistry and pharmacology Arika Focus. **2010**, *23*(1), 53–83.

89. Bhattacharyya, P. N.; and Jha, D. K.; Antidermatophytic and antioxidant activity of *Terminalia arjuna* (roxb.) Wight & Arn. Bark *Int. J. Pharm. Biol. Archives.* **2011**, *2*(3), 973–979.

90. Khan, Z. M. H.; Faruquee, H.; and Shaik, M.; Phytochemistry and pharmacological potential of *Terminalia arjuna* L. *Med. Plant. Res.* **2013**, *3*(10), 70–77.

91. Mota, K. S. L.; Pereira, F. O.; Oliveira, W. A.; Lima, I. O.; and Lima, E. O.; Antifungal acitivity of Thymus vulgaris L. Essential oil and its constituent phytochemicals against Rhizopus oryzae: Interaction with ergosterol Molecules, **2012**, *17*, 14414–14433.

92. Hudaib, M.; Speroni, E.; Pietra, A. M.; and Cavrini, V.; GC/MS evaluation of thyme (*Thymus vulgaris* L.) oil composition and variations during the vegetative cycle. *J. Pharm. Biomed. Anal.* **2002**, *29*, 691–700.

93. Bindu, S.; Padma, K.; and Suresh, C. J.; Topical treatment of dermatophytic lesion on mice (Mus musculus) model. *Ind. J. Microbiol.* **2011**, *51*(2), 217–222.

94. Kuldeep, G.; and Pathak, A. K.; Pharmacognostic and phytochemical evaluation of *Tridax procumbens* Linn. *J. Pharmacognosy. Phytochem.* **2013**, *1*(5), 42–46.

95. Kannan, P.; Shanmugavadivu, B.; Petchiammal, C.; Hopper, W. *In vitro* antimicrobial activity against skin microorganisms. Acta Botanica Hungarica **2006**, *48*(3–4), 323–329.

96. Ponnusamy, K.; Chelladura, P.; Ramasamy, M.; and Waheeta, H.; *In vitro* antifungal activity of indirubin isolated from a South Indian ethnomedicinal plant *Wrightia tinctoria* R. Br. J. Ethnopharmacol. **2010**, *132*, 349–354.

97. Ranjani, M.; Deepa, S.; Kalaivani, K.; and Sheela, P.; Antibacterial and antifungal screening of ethanol leaf extract of *Wrightia tinctoria* against some pathogenic microorganisms. *Drug Invention Today,* **2012**, *4*(5), 365–367.

98. Anusharaj, Chandrashekar, R.; Prabhakar, A.; Rao, S. N.; Santanusaha. *Wrightia tinctoria*: An overview. *J. Drug. Deliv. Therap.* **2013**, *3*(2), 196–198.

99. Meenakshi, S.; and Richa, S.; Synergistic Antifungal Activity of *Curcuma longa* (Turmeric) and *Zingiber officinale* (Ginger) Essential Oils Against Dermatophyte Infections. *J. Essen. Oil. Bear. Plants.* **2011**, *14*(1), 38–47.

100. Shenq-ji, P.; Ethnobotanical approaches of traditional medicine studies some experiences from Asia. *Pharam. Biol.* **2001**, *39*, 74–79.

CHAPTER 3

A NOTE ON AN INTEGRATED HOLISTIC APPROACH FOR CANCER PAIN

M. SHARANYA[1] and D. R. SATHISHKUMAR[2]

[1]Department of Bioinformatics, Bharathiar University, Coimbatore - 641 046, India

[2]Department of Biotechnology, Salem Sowdeswari College, Salem - 636 010, India

CONTENTS

3.1   INTRODUCTION

- What is pain?
  Pain is a unpleasant disturbed_sensation which accompanies the activation of nociceptors.

MECHANISM OF PAIN SENSATION:

Nociceptors are which carries pain stimulus.
- Any physical, chemical, thermal or mechanical stimulus like heat, cold, pressure activates these nociceptors.
- These are free nerve endings found in all body tissues.
- They carry pain stimulus to higher centers.
- Once a nociceptors is stimulated, it releases a neuropeptide, which initiates electrical impulse along the afferent fibers toward the spinal cord.
- These afferent fibers are of two types:
  - A delta fiber
  - C delta fibers

TYPES OF PAIN:

1. Acute pain
2. Chronic pain
3. Somatic or visceral pain
4. Referred pain
5. Emotional or psychogenic pain

Acute Pain: It is a warning. It seems obvious that pain enables an organism to sense tissue damage and avoid further damage, thus facilitating survival. The damage to tissues by mechanical, thermal, chemical agents, directly activates neurons signal in the form of noxious stimuli, the initial phasic stage of pain takes on a tonic persistence until healing can occur. Acute pain can be thought of as both a sensation of actual or impending tissue damage as an indication for rest.

Chronic Pain: It is a disease. This pain sensation persist beyond the period of time required for healing and can become very destructive because of the negative physiological, psychological, and social consequences to the individual. Pain doesn't serve a positive role in triggering escape or withdrawal behavior to prevent further harm nor does it serve to promote recovery. This pain develops out of acute pain appears insidiously. In chronic pain there is change in the pattern of sympathetic and adrenal mechanism. There appears to b habituation of the sympathetic responses to moderate pain with the appearance of vegetative sign. Individuals show delayed sleep onset and frequent awakening. Pain tolerance is lowered, thus with additional noxious stimuli, individual may appear to overreact in a given stimulation.

The individual adopts the role of a sick and behavior of a chronic invalid, which are extremely difficult to extinguish or modify. There may be central mechanism that enhances the vegetative behavior associated with the chronic pain. It is defined as pain that continues after the stimulus is removed or the damaged tissues heals. Physiologically, chronic pain is believed to be result from hyper sensitization of the pain receptors and enlargement of the receptor field in response to the localized inflammation that follows the tissue damage. This pain is poorly localized, has an ill defined time of onset and is strongly associated with subjective components. The effects of pain experienced extent beyond the individual and affects the family, the work place and the social sphere of the individual.

Somatic and visceral pain: A distinction must be made between pain arising from the body and that from the viscera. Pain arising from the body wall may be superficial or deep. Superficial pain arises from the skin and is localized. Deep pain arises from the muscles, joints, fascia, tendons and periostum. It tends to be more diffuse but can be localized. It may last for longer periods of time and is often accompanied by muscle spasms, which are themselves a source of pain. Visceral pain arises from organs in abdominal and thoracic cavities either from visceral or the parietal muscles.

Referred pain: The pain initiated in deep body structures may be localized at sites some distance from the actual location of pathological origin is called referred pain. The mechanism which account for this referral of pain are based on the anatomical arrangement of dermatomes, sclerotomes and myotomes and the convergence of the cutaneous and visceral afferents with the spinal cord.

Emotional or psychogenic pain: Evidence for emotionally induced pain doesn't extend beyond clinicians reports. Chronically anxious and depressed people appear to be vulnerable to pain. Fear, anxiety and depression are capable of amplifying pain stress can give rise to or amplify pain. Anxiety, depression, anger and aggressive behavior may provoke substantial automatic, visceral and skeletal activity which can enhance pain sensation. Patients often focus on the somatic evidence of pain and are reluctant to discuss underlying psychological stress, making accurate diagnosis of the cause of pain and successful treatment extremely difficult.

3.2   HOLISTIC APPROACH

By definition, the word "holistic" means: "relating to or concerned with wholes or complete systems rather than with the analysis of, treatment of, or dissection into parts—medicine which attempts to treat both the mind and the body." As such, true holistic therapies must include more than just a set of symptoms to be addressed. They must consider the whole being, whether human or animal, including the emotional, mental, and physical (both internal and external) environments in which they exist. Some examples of holistic approaches include Homeopathy, Acupuncture, Ayurveda, etc.

The modern medical model largely views the body as a collection of pieces and parts to be assessed in isolation rather than as a working unit. Symptoms are viewed as disease, and treatment is aimed at suppressing rather than eliminating those symptoms. In the holistic view, symptoms are regarded as the body's expression of imbalance and an attempt to return to homeostasis (a state of equilibrium).

HOLISTIC APPROACHES FOR CHRONIC PAIN:

1. Acupuncture
2. Massage therapy
3. Yoga therapy
4. Tai Chi
5. Ayurveda
6. Homeopathy
7. Biofeedback
8. Physical therapy
9. Mind body technique
10. Reiki healing
11. Hypnosis
12. Meditation

1. Acupuncture: Acupuncture is one of the oldest and most commonly used holistic medical approach in the world. The word acupuncture describes a variety of procedures involving stimulation of anatomical points on the body by using variety of techniques. Acupuncture points are believed to be points that allow entry into channels. This is to redirect, increase or decrease the body's vital substance, qi (produced chi) and restore balance on an emotional, spiritual and physical level.

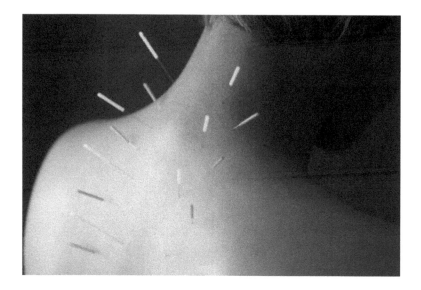

2. Massage Therapy: Massage is one of the most effective healing arts. Gentle massage affects the nervous system through nerve endings in the skin stimulating the release of endorphins, the body's natural "feel good" chemicals to help induce relaxation and a sense of well-being to relieve pain and reduce levels of stress chemicals. It helps reverse the damaging effects of stress by slowing Heart rate, Respiratory rate and Metabolism and lowering raised Blood pressure.

Stronger massage stimulates blood circulation to improve supply of oxygen and nutrients to the body tissue and helps the lymphatic system to flush away waste products. It eases tense, knotted muscles and stiff joints, improving mobility and flexibility. Its many benefits includes deep relaxation, improve circulation and postural awareness.

The holistic approach offers:
(a) Aromatherapy massage
(b) Thai massage
(c) Swedish massage
(d) Transforming touch massage

a) Aromatherapy massage: It is a gentle treatment using plant oils together with massage, a powerful, soothing way to support the body on many levels, wonderful for reducing the negative effects of stress.

b) Thai massage: It is a combination of Yoga, Massage and Acupressure techniques. It is performed on floor mates with the patient fully clothed. The recipient will experience a deep sense of calm, improved flexibility followed by renewed energy and vitality.

c) Swedish massage: It can be performed as a whole body treatment or to target specific areas of tension ,it is an invigorating and rejuvenating treatment ideally suited to more physical alignments.

d) Transforming touch massage: It works on body mind and spirit. The beautiful treatment relaxation to a new level. Transforming touch is ideal for everyone, it is gentle yet deep, soothing yet recharging.

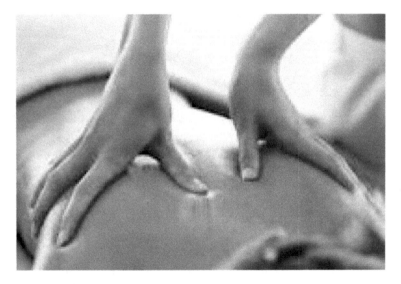

3. Yoga therapy: Yoga is the union of individual 'Self, 'Jeevatma', with the universal self, 'paramatma'. It is the communication of human soul with the divinity. The essence of our existence is our individual principle of Consciousness or the simple meaning of YOGA is union. Yoga is the integration and harmony between thoughts, words, and deeds or integration between head, heart, hands. Aasanas that helps to cure disease and symptoms.

Sputa Baddha Konassana, Garudasana, Anantasana, Supta Virasana, Salbhasan, Bhujangasana,Ustarasana, Urdha Mukha Svaasana, and Dhanurasana have effects on arthritis, backache, and spondylitis. Supta Virasana, Adhomukha Virasana, Pawanmuktasana, Halasana, Bhujangasana, Bharadvajasana, Janu sirsasana, Pascimmotasana, and Nadi Sodhana Pranayam have effects on diabetes and diabetic pain.

4. Tai Chi: Tai chi is a mind body practice in complementary and alternative medicine that originated in China as a form of martial art. Often referred to as "moving meditation," tai chi practitioners move their bodies slowly, gently, and with awareness, while breathing deeply. People practice tai chi for various health-related purposes, such as to improve physical condition, muscle strength, coordination, and flexibility. It can also be practiced to improve balance and decrease the risk of falls, especially in elderly people, as well as to ease pain and stiffness.

5. Ayurveda: Ayurveda is an indigenous ethnic medical system in popular practice in the Indian subcontinental since the pre-biblical era. The system's core strength is its holistic approach to health and disease using natural remedies derived from medicinal plants and minerals. It is a complete and a series of wellness treatments to find balance and health within the body can be beneficial to those already suffering from aliments such as chronic pain.

Ayurveda is more than a mere healing system, it is a science and art of pain defense formula. Ayurveda includes 4 basic keys to enhance and empower your body, rejuvenating the body's functions.

1, Systematic relaxation 2, Massage 3, Diet 4, Gentle asana

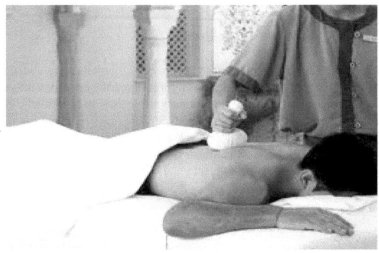

6. Homeopathy: It is a use of infinitesimal doses of substances to influence bodily remedies in which the medication contains several homeopathic remedies. Homeopathic pain relief is brought about through its medicines derived from minerals, plants, and animals.

7. Biofeedback: Biofeedback is based on the concept of using your mind to improve your health. Some health professionals believe that individuals have the ability to influence their thoughts to control some of the body's involuntary functions. Special monitoring equipment is used to teach control of certain body functions and their responses.

8. Physical therapy: Physiotherapy works to improve mobility and health and to reduce the risk of injuries. An injury caused by a range of factors working together. Physiotherapy approach aims to lessen the risk of the injury happening again. Physiotherapy have wide range of therapies. Some of these therapy include:

Manual therapy: as massage, stretching, manipulation, manual resistance.

Electrotherapy techniques: Ultrasound, TENS, LASER, and diathermy.

Exercise program: posture retraining, stretching, muscle strengthening.

Other services: tapping, splinting.

9. Hypnosis: It is an altered state of awareness used by therapists to treat psychological or physical problems. During hypnosis, the conscious part of the brain is temporarily tuned out as the person focuses on relaxation and lets go of distracting thoughts. In hypnosis, hypnotists use a magnifying glass to focus the rays of the sun and make them more powerful. When our minds are concentrated and focused, we are able to use them more powerfully. When hypnotized, a person may experience physiologic changes, such as a slowing of the pulse and respiration, and an increase in alpha brain waves.

10. Meditation: Meditation involves using a number of awareness techniques to help quiet the mind and relax the body. The two most common techniques are:

(a) Transcendental meditation: The patient repeats a single word or phrase, called a mantra, and is taught to allow other thoughts and feelings to pass.

(b) Mindfulness Meditation: The person focuses all of his or her attention on thoughts and sensations. This form of meditation is often taught in stress-reduction programs.

11. Reiki healing: Reiki is a healing practice that works directly with the energy of the body. Enhancing the body's own ability to heal its all or tends to heal physical, emotional, mental, and spiritual problems. It is a hands on healing technique. Reiki allow the body to relax and move into a calm and meditative. This allows the body's energy field to reenergize, release heaviness and bring in a new lighten, a sense of peace and of deep relaxation. The effect of reiki can last for several weeks. Reiki is a painless, noninvasive treatment that does not interfere with medication.

12. Mind body techniques: psychological factors are important contribution to intensity of pain and to the disability associated with chronic pain. Pain and stress are intimately related. There may be a vicious cycle in which pain cause stress and stress is then cause more pain. These approaches provide a variety of benefits including a great sense of control, decreased pain intensity and distress, change in the way pain is perceived and understood, and an increased sense of well-being and relaxation. These approaches may be valuable for adult and elder with pain.

3.2.1 PHYSIOTHERAPY APPROACH FOR CHRONIC CANCER PAIN

American Pain Society quality of care task force for treatment of acute and chronic cancer pain had recommended relaxation, heat, cold, deep breathing, walking, imagery or visualization under nonpharmacological methods for cancer pain relief.

Thus, the physical therapists have a very important role to play in holistic care of patients diagnosed with cancer as stated by Flomenhoft and Rashleigh. Rashleigh listed the therapeutic strategies employed by physical therapists in palliative oncology as ambulation and musculoskeletal therapy, neurological therapy, respiratory therapy, electrophysical agents, mechanical therapy, decongestive physiotherapy, and education. Santiago-Palma and Payne listed treatments used by physical therapists on cancer patients as therapeutic massage, therapeutic heat, therapeutic cold, patient education (advice on activity modification), range of motion and strengthening exercise, training ambulation using assistive devices, environmental modification, energy conservation and work simplification techniques. Twycross mentioned that physical treatment methods like massage, heat pads and TENS are useful for pain management in cancer patients.

Physical therapy treatment techniques have also been reported in cancer-related fatigue by Watson and Mock, in breast cancer, prostate cancer and breast cancer-related lymphedema, older women with cancer, cancer therapy-related hyperthermia and colorectal cancer.

The physical therapy treatment modalities and methods are listed here building upon existing evidence under the five mechanism-based classification categories of cancer pain.

(i) Central sensitization mechanism-based physical therapy for cancer pain:

Bennett *et al*, in their detailed systematic review with meta-analysis of 15 studies found that educational interventions (written and/or audiovisual learning materials) improved knowledge and attitudes toward cancer pain and analgesia, and perceived pain intensity among cancer patients. Pain educational programs were shown to be highly effective not only in reducing pain and associated pain behaviors, but also in reducing treatment-related barriers in cancer patients. One such method is the use of pain management diary.

Evidence for using TENS for pain relief in cancer patients is continuously growing. TENS addresses the central component of cancer pain and is a very useful therapeutic adjunct in patients with central sensitization.

(ii) Peripheral sensitization mechanism-based physical therapy for cancer pain:

Regardless of the patient's prognosis, rehabilitation for neuropathic pain in cancer patients may enhance function, and attention to safety factors may avoid serious accidents.

Rehabilitation of patients with motor deficits on neurological examination begins with assessment of the patient's functional dependence—their ability to walk, dress, prepare meals, and perform other activities. Assistive devices may be useful when there is impairment in any of these activities. Physical therapy can increase the strength of involved muscles as well as accessory muscles, which can improve coordination and sensory integration. Physical activity also maintains muscle and ligament length, preventing later deformities. Ankle foot orthotics (AFO)-type braces, which fit easily within a standard shoe, can help prevent falls when patients experience a slapping gait or foot drop.

Using neurodynamic testing on patients with "nerve trunk pain" (stimulus-evoked pain along the course of the nerve) and nerve mobilization (neural manual therapy technique as described by Kumar and Jim and nerve massage is useful therapeutic adjuncts.

Hypersensitive skin can be treated with desensitization measures with other alternative sensory stimuli that are tolerated fairly by the patient. A hyposensitive skin can be treated with sensory reeducation methods using various forms and textures of materials. A variety of physiotherapy treatment methods like electrical simulation, magnetic therapy, pulsed electromagnetic energy, photon stimulation, monochromatic near infrared therapy for peripheral neuropathic pain are used.

(iii) Sympathetically maintained mechanism-based physical therapy for cancer pain:

Cold therapy may reduce swelling and relieve pain longer than heat therapy by decreasing nerve conduction velocity and desensitization of free nerve endings of the skin. This method can either be utilized at the painful region itself or as remote desensitization. Though studies on physical therapy management for sympathetic pain in cancer do not exist, management in complex regional pain syndrome (CRPS) and other altered sympathetic states includes TENS (burst mode) application to the related spinal level, combined with relaxation and biofeedback techniques to restore vasomotor balance.

One of the manual therapy techniques (neural manual therapy technique as described by Kumar and Jim commonly employed by physical therapists was the "sympathetic slump" or "slump long sitting with sympathetic emphasis", a sympathetic nervous system mobilization technique, which was shown to have sympatho-exicitatory effects in extremities including sudomotor and vasomotor effects. The technique is very simple and is well tolerated by patients and was shown to be useful in CRPS patients. Another technique is the use of thoracic spine mobilization in patients who had spinal dysfunction which is another useful manual physical therapy method.

(iv) Nociceptive-mechanism-based physical therapy for cancer pain:

Massage therapy improves local circulation and gently stimulates the free nerve endings, the pressure of which may also help in draining local tissue edema and induce local and general relaxation. One of the well-established scientific forms of massage is the manual lymphatic drainage therapy and complete decongestive therapy (manual lymphatic drainage, compression garments, skin care and range of motion exercises). Massage therapy was shown to be very effective to relieve symptoms of cancer pain in numerous studies.

Reeves explained the importance of physical interventions such as changes in patient positioning, relaxation techniques for sleeplessness, and energy conservation techniques for fatigue in patients with cancer pain. Therapeutic modalities, such as electrical stimulation (including transcutaneous electrical neurostimulation), heat, or cryotherapy, can be useful adjuncts to standard analgesic therapy in patients with cancer-treatment-related lymphedema and pain. The treatment of lymphedema by use of wraps, pressure stockings, or pneumatic pump devices may both improve function and relieve pain and heaviness.

Mufazalov and Gazizov showed that laser therapy enhanced therapeutic efficacy of pain-relieving drug regimen in patients with cancer pain. Cancer treatments like radiation therapy can induce mucositis in patients with oral or head and neck cancer and can cause oral pain due to impaired wound healing. Bensadoun commented on the importance of low-level laser therapy on wound healing and its role in mucositis treatments. Maiya *et al*, subsequently showed that helium–neon laser therapy was effective to reduce pain and improve healing of radiation-induced mucositis after 6 weeks of therapy in head and neck cancer patients.

The benefits of exercise and increased physical activity on people diagnosed with cancer are many, including improved function, quality of life, strength, and endurance, and reduced depression, nausea, and pain. Beaton *et al*, in their systematic review found strong, high-quality evidence in favor of exercise interventions (aerobic exercises and strength training given alone or as part of a multimodal physical therapy intervention) in patients with metastatic cancer for improving physical and quality of life measures. McNeely *et al*, found that progressive resisted exercise training (PRET) program significantly reduced shoulder pain and disability and improved upper extremity muscular strength and endurance in postsurgical head and neck cancer survivors who had shoulder dysfunction because of spinal accessory nerve damage.

Keays *et al*, found improvements in shoulder range of motion and function in women with breast cancer undergoing radiation therapy, who were given pilates exercises which involves whole body movements with breath control. Similar improvements in pain and mobility were observed following physiotherapy intervention (exercises, advice, soft tissue massage to surgical scar) in breast cancer patients who underwent axillary dissection.

Graded activity prescription and regular physical activity as a component of multimodal approach in treatment of cancer pain have a direct influence on the peripheral musculoskeletal system via the exercising muscles. Regular physical activity bears a direct effect on tissue functions, thus leading to counter irritation phenomenon of pain relief. However, for these programs to be effective, it should be accompanied with behavioral training and patient education.

(v) Cognitive-affective mechanism-based physical therapy for cancer pain:

Cognitive behavioral therapy (CBT) intervention comprising education, distraction, relaxation, positive mood development and self-coping strategies for pain helps the patient develop acceptance toward symptom persistence and enables them to lead a functionally active life. CBT was studied extensively and shown to be effective in a large number of studies.

Music therapy for pain relief in cancer patients was shown to be effective by many authors across the globe.

KEYWORDS

- **Complex regional pain syndrome (CRPS)**
- **Chronic pain**
- **Holistic approach**
- **Rehabilitation**

CHAPTER 4

SYSTEM FOR COMPLEX EXPRESS ANALYSIS OF MICROBICAL INFECTION, AS WELL AS BIOTESTING OF VARIOUS MEDIA, PRODUCTS, AND PREPARATIONS

V. S. SIBIRTSEV

GiproRjibFlot (Research and design institute on development and exploitation of a fish fleet), lab. Technical microbiology; Instrumentalnaja ul. 8, St. Petersburg, 197022 Russia; E-mail: vs1969r@mail.ru

CONTENTS

At different stages of manufacture and quality surveillance of food, medicinal and etc., as well as at valuation of sanitary–epidemic state of water, air, working and residential zones, etc., more and more necessary becomes the introduction as it is possible for more fast and simple methods of microbical infection valuation.

For this purpose we develop a agreed system of express–analysis, including valuation of common contents Adenosine 5′-triphosphate (ATP) in samples (*on ATP inducible chemifluorescention at the presence of specific enzyme*), peptides, and nucleic acids (*on photofluorescention of its in presence specific dyes*) with the subsequent recalculation of these parameters on quantity of microbical cells in sample.

In aggregate, these methods give the large information, than on separateness. In particular, under the contents ATP it is possible to evaluate availability in sample only live cells. The contents of peptides and nucleotides outside of active cells can testify to availability in sample viruses, as well as inactivate cells, its fragments, and other potential dangerous for people factors. By system from several fluorescent dyes with understood spectral and binding properties, it is possible to evaluate not only quantity, but also structure (*including space— which varies depending on a condition of cells even directly during their live*) of DNA, RNA, and peptides in researched samples—as is described, in particular, in work [1–4]. And by appropriate selection detergents (*making those or other types of cell membranes and walls accessible for passage through them of used «external» dyes*), it is possible to do accessible for analysis only determined types of cells.

Thus, in difference from «standard cultural (*growth*)» techniques (*requiring for its realization from several days up to several weeks and conditions, provided only in specialized microbiological laboratories; as well as considerably dependent from qualification and experience of the employees, performing these analyses*), such analysis, if necessary, can be carried out directly on a place of selection of probes, during several minutes, at the help of a simple, portable, cordless, automatic devices.

Thus, selection of samples from air and other gas media is made at the help of a specialized (*but also cordless*) aspirators (*on filters with subsequent their washed or Petri dish with subsequent its incubated*). The analysis of large volume liquid samples is also executed by them filtration (*passive or with application of pumps*). The analysis of firm samples is made for its homogenization (*including with the help appropriate, specialized, cordless devices*) with appropriate liquid media. Analysis of firm surfaces is executed by selection smears from the determined area of these surfaces or washed with its.

And in case of necessity of specify of species structure of microbical infection, the probes can be in addition analyses in conditions of specialized microbiological laboratory with application as «standard cultural (*growth*)» techniques, as its updating, using immune-enzyme analysis, impedance microbical technology (*on change electropassage of selective mediums, occurring owing to live of analyzed microbes in these mediums*), PCR-diagnostics, etc.

Besides recently, more and more urgent becomes the problem of fast revealing all possible as positive, as negative properties of new food products, biology active substances, medicinal, and other preparations. And not only at isolated its action on organism, but also in a combination to other products, preparations, and environment factors, specific for various regions (*stress, electromagnetic, and acoustic noises*), increased radiating background, duration of a light day, humidity, dust, etc.

We for such valuation use testy systems—as which are used as microorganisms, as microorganisms (*plants, fishes, rats, etc.*).

Thus microorganisms once or sometimes through certain periods enter a analyzed product or preparation. Then at each of such organisms regularly select small samples of fabrics or blood. And then evaluate (*with the help of a system-specific fluorescent dyes which understood spectral and binding properties*) change of a genome structure in cells of this samples in comparison with control group.

At the same time, for microorganisms and a number of small microorganisms, besides usual valuation of this survival as well as changes in a structure of cell genome are analyzed, in particular, curb the growth and vital activity of these organisms at the presence of tested preparations—on change of electropass property of medium, in which it lives. Thus, for the last impedance analyzer, «Rabit» is used, enabling in a real mode of a time to register change of electropass property of selective medium, owing to vital activity in its analyzed organisms simultaneously in many samples (*up to several hundreds*), with a separately set time and incubation temperatures, as well as other parameters.

KEYWORDS

- **Biotesting**
- **Microbical express–analysis**
- **DNA–protein–ATP fluorescent analysis**
- **Impedance microbical technology**

REFERENCES

1. Ivanov S. D.; and Sibirtsev, V. S.; Method of valuation general microbiological pollution. Russia patent №2079138, bulletin №13 (1997). (in Russian).
2. Sibirtsev, V. S.; Biochemistry (Moscow). **2005**, *70*(4), 449–457.
3. Sibirtsev, V. S.; and Garabadgiu, A. V.; Fluorescent DNA-probes: introduction in the theory and practice of use. NOU Express. Saint-Petersburg, **2006**. 188 p. (in Russian).
4. Sibirtsev, V. S.; and Garabadgiu, A. V.; *Biotechnosphere 13*(3), **2011**, 9–14, (in Russian).

CHAPTER 5

A STUDY ON THE INFLUENCE OF THE TREATMENT BY THE CLEANED SOLUBLE PROTEINS TO THE FRAGMENTED SARCOPLASMIC RETICULUM

O. M. ALEKSEEVA

Emanuel Institute of Biochemical Physics, Russian Academy of Sciences, Moscow, ul. Kosygina, 4, Moscow, Russia, 119334; Fax: (499) 137-41-01; E-mail: olgavek@yandex.ru

CONTENTS

5.1   INTRODUCTION

The basic calcium store of striated muscles of mammals are the sarcoplasmic reticulum (SR). The SR is the branched tubules and vacuoles that are linked with each other between them into the continual net inside of the muscle cell interior. It exists from cell periphery up to central compartments. The peripheral parts of reticulum are the terminal cysternaes (TC). They are spaced at 40 Å from plasmalemma, and are mechanically connected by means of long proteins with potential-depended channels at the cell surface. The longitudinal tubules exist at the muscle cell interior at middle parts. So, that all contractile apparatus acto-miozine complex of striated muscle were braided of SR net (Figure 5.1).

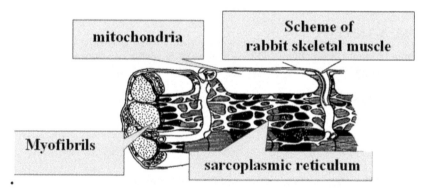

FIGURE 5.1   Scheme of rabbit skeletal muscle.

For the initiation of contractile apparatus activity the bursts of the calcium ions concentration are need obligatory. The relaxations of muscle are needed by calcium ions concentration lowering. The net of SR is exercising both functions. It activates the muscle contraction, increasing the concentration of calcium ions, when release of ions from TC occurred. The relaxation of the muscle begins, when pumped the ions in to inside lumen of reticulum store. The calcium release occurred by the ryanodine receptor (RyR) activation. And pumping of calcium ions occurred by the Ca^{2+}-Mg^{2+}-dependent ATPase (SERCA2) activation. SERCA2 is located previously at the longitudinal tubules, and smaller portions of SERCA are located at the terminal tubules for the maintaining of Ca^{2+}-gradient through the SR membrane. RyR is located at the terminal cistern only. It is exhibited apart to the junction space between SR and plasmalemma (Figure 5.2).

Our work deals with two model objects that imitated the SR structure and activity. The two fractions of fragmented reticulum consist of the suspension of isolated membrane bubbles. Fragmented SR (FSR) was prepared: from the terminal cistern—the heavy caffeine-sensitive fraction. From the longitudinal tubules we prepared the light fraction that was separated into the 2 subfractions: caffeine-sensitive and caffeine-insensitive. The ryanodine receptor (RyR) is the main Ca^{2+}-releasing channel. RyR's activity is regulated by Ca^{2+} and a numerous of the endogenous and the pharmacological ligands (RyR was blocked or activated by the plant alkaloid—ryanodine, and only was activated by methylksantin—caffeine and its agonists [1].

The luminal, transmembrane and cytoplasmic satellite proteins: calsequestrin, triadin, janctin and others, modulate the function of RyR also under the certain values of the extra vesicular—cytoplasm ($[Ca^{2+}]_{cyt}$) and the luminal Ca^{2+}-concentrations ($[Ca^{2+}]_{lum}$). The necessary values of $[Ca^{2+}]_{cyt}$, $[Ca^{2+}]_{lum}$ are based on the functional activity of the RyR and Ca^{2+}-pump - Ca^{2+}-ATPase. The gradient of Ca^{2+}—$[Ca^{2+}]_{lum}/[Ca^{2+}]_{cyt}$ is created and maintained by coordination of Ca^{2+}-ATPase and RyR activities. Data [2] suggested, that $[Ca^{2+}]_{lum}/[Ca^{2+}]_{cyt}$ may be the key factor, involved in the regulation of the functional states of Ca^{2+}-ATPase and RyR of Ca^{2+}-store—SR. The maximal value of the calcium loading of the Ca^{2+}-store (3mM) induced the Ca^{2+} -releasing process through the Ca^{2+} -channel of RyR, inducing the muscle contraction. In this case the Ca^{2+}-ATPase is inhibited completely, and no pumps the amount of the released Ca^{2+} to the reticular lumen. Minimal Ca^{2+}-concentration at the reticular lumen (0–0.3 mM) completely inhibits the Ca^{2+} release through the RyR, and activates the Ca^{2+}-ATPase, resulting in the Ca^{2+}-reuptake from the cytoplasm, and followed by muscle relaxation. The middle Ca^{2+}-concentration (1 mM) corresponds with the equilibrium position of the system Ca^{2+}-releasing/Ca^{2+}-pumping, when RyR and Ca^{2+}-ATPase may be activated both one after other [3].

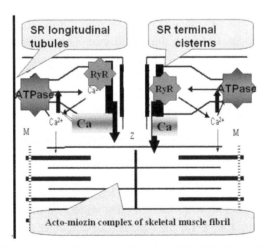

FIGURE 5.2   Scheme of sarcoplasmic reticulum in rabbit skeletal muscle.

The efficiency of Ca^{2+}-accumulation (Ca^{2+}/ATP) by heavy FSR, and light FSR was recorded by the registration of the of incubation medium acidifications. These measurements we made, when we used Mg-ATP for Ca^{2+} ions pumping inside to the FSR lumen. The value of Ca^{2+}/ATP reflects the amount of pumped Ca^{2+} ions to one molecule of hydrolyzed ATP.

The bilayer integrity of membrane provides the efficient operation of calcium stores. However when pathologies, in the process of aging of membrane, when action of damaging factors, the regular structure of bilayer breaks down and the passive leakage of ions occurred, without any related work of specific channels. This is why it was necessary to show that when extraction from reticular membranes of free fatty acids (FFA) occurred, which were accumulated in membrane at above case, the passive leakage decreased, that promoted the successful work store.

5.2   MATERIALS AND METHODS

The materials: KCl, KH_2PO_4 (Merck); histidine, imidasol, caffeine (Merck); NaCl, $MgCl_2$ (Merck); DTT (Serva); glycerol (Serva); $CaCl_2$ (Merck); sucrose (Merck); EGTA (Serva); PMSF (Helicon); HSA (Sigma).

The standard methods of isolation and the purification of FSR were modified. The first step was realized in the presence of DTT, PMSF and 10 mM caffeine, and with addition of aggregation stage in glycerin medium at final step [4–6]. FSR were prepared from white muscles back legs of rabbit. Muscle was cooled in physiological solution, and was crushed on meat grinder. Crushed muscles 200 g had placed in 600 ml of medium, containing 0,3 M sucrose, 10 mM caffeine and 10 mM histidine (pH 7,7 4°C). Then it was homogenized at a temperature of 2–4°C by means of homogenizer "Politron". The homogenate was centrifuged under 10 000 g during 20 min. The supernatant was pelleted again by centrifugation 36 000 g during 60 min. The pelleted total fraction of membranes was extracted 60 min in the cold medium, containing: 0,6 M KCl, 0,1mM EDTA, 0,2 mM $CaCl_2$, HSA (0,6mg/ml), 5 mM histidine (pH 7,4 4°C). The suspension again centrifuged 11 000 g for the pelleted of fragments of TC. Then supernatant pelleted by centrifugation 40 000 g during 60 min for the deposition of fragments of longitudinal tubules (LT). The obtained pellets in practice not contained of heavy mitochondrial fragments. For FSR holding pellets were suspended at storage medium, contained 25 percent glycerin (in volume of), 0,1mM EDTA, 0,2 mM $CaCl_2$, 5 mM histidine (pH 7,4 4°C). For following cleaning of factions, it were layered on storage medium and centrifuged 36 000 g during 60 min. Pellet was the fragments, responsive to caffeine (TC), low layer of suspension, and light fraction (LT). Protein content was standard for the

FSR: TC contained the RyR, Ca^{2+} -ATPase and calsequestrin predominantly. LT contained the Ca^{2+} -ATPase predominantly and lesser amount of calsequestrin [7].

The protein concentration was determined by the fast method [8].

Efficiency of Ca^{2+}-accumulation by heavy FSR, and light FSR was recorded by potentiometric method (pH-metric) [9]. FSR vesicles (3–4 mkg/ml) were incubated in 4 ml medium, contained 2 mM ATP, 5mM sodium oxalate, 0,1M NaCl, 4 mM $MgCl_2$, 2,5 mM imidasol (pH 6,8 37°C) with intensive mixing. The pumping reaction was stimulated by additions of 80 nmoles $CaCl_2$. The Ca^{2+} ions, accumulated into FSR, bounded with oxalate, and the calcium oxalate stored into the SR lumen. Thus capacitance FSR for Ca^{2+}-ions increased markedly. It becomes to perform the registration of absorption Ca^{2+}-ions by FSR eventually. In native muscle cells under the SR activation, Ca^{2+}-ions communicate with luminal proteins, calsequestrin and others, and with phosphate. So that capacitance SR for Ca^{2+}-ions was maintained. The value of Ca^{2+}-ions saturating of Ca^{2+}-binding proteins is the essential regulator factor for RyR activity.

The loading FSR by the calcium oxalate into the SR lumen carried out in 40 ml of medium that used for the activity registration in standard conditions. The additions of $CaCl_2$ were held as Ca^{2+}-ions was accumulated into the FSR lumen. Reticulum (5 mg of protein) was loaded up to 500 nmol Ca^{2+}/1 mg protein. Then the mixture has cooled up to 4°C and centrifuged during 1 hour. The FSR pellet was washed out on medium, contained: 2 mM ATP, 0,1 M NaCl, 5 mM Na oxalate, 0,5 mM imidasol (pH 7,0 4°C). Then the loaded FSR was suspended in 1 ml of storage medium, contained 0,1 M NaCl, 0,5 mM imidasol (pH 7,0 4°C).

The measurement of Ca^{2+} passive release was held by the pH-registration method [9], at incubation medium, contained: 0,1 M NaCl, 0,5 mM EGTA, 0,5 mM imidasol (pH 7,0 37–42°C). The Mg^{2+} concentration varied from 0,1 mM up to 10 mM.

The cleaning of human serum albumin (HSA) from calcium and FFA carried out by next method. Albumin 1g was solubilized in distilled water 5 ml, pH up to 3, 0–3, 5 (on ice).The pharmaceutical activated charcoal was added to albumin-water solution. This mixture was incubated 1 hr under the constant mixing. Then the charcoal was pelleted by centrifugation or filtering. And pH-value of supernatant (that contained of cleaned albumin) was led up to physiology (7,0). The dialysis against solution EDTA 10^{-4} M was held the daily on ice. The change of the solution occurred a few times. The cleaned albumin solution kept in frozen condition.

The cleaning FSR from FFA was held by means of cleaned albumin by the incubation way of membranes FSR at medium for the Ca^{2+}-transport registration, containing 2–5 mg FSR protein and 10–20 mg /ml of HSA, refined from fatty acids, and 1,9 mM ATP during 2 min (25–30°C). Then mixture was cooled centrifuged 40 000 g 60 min. The pellet was suspended in storage medium. The FSR cleanness was controlled by registries the Ca^{2+} transport at medium with low concentration of oxalate (1,5 mM) with and without albumin.

The extraction of lipid from FSR was made by Folch method [80], [10] 20–25 mg of FSR protein in the 2 ml was homogenized by homogenizator "Politron" under the ice cooling. The medium (40 ml) for extraction consist of 2 part of chloroform, 1 part of methanol and antioxidant ionol (4 metil-2,6-ditertbutilphenol) (1mg/l). Homogenate was filtrated. Then 10 ml 0,1 M KCl were added. And this mixture was homogenized during 10 min. Homogenate was centrifuged 25,000 g during 40 min. Then the undersize of phospholipids in chloroform was being selected and was evaporating. The phospholipids were solubilized with methanol-geptan mixture (4: 1)

The building of FFA in FSR membrane is carried out by next procedures. The ethanol solution of free fatty acid was added to FSR membranes, suspended in storage medium (15–25 mg FFA/1 mg FSR protein). This mixture was incubated during 3–4 hr (10–12°C) for more uniform distribution of fatty acid in FSR membrane

5.3   RESULTS AND DISCUSSION

As the Ca^{2+}-gradient play the great role for the SR functions regulation, we were needed to investigate some factors influences to the Ca^{2+}-store actions. One of these factors—is the membrane permeability for ions. Because, the Ca^{2+}-ions penetrate through the membrane by several ways, and Ca^{2+}-pumping by SER-CA2 activity, and Ca^{2+}-releasing by RyR activity are under widespread intensive investigations, we reversed our attention on the nonspecific way passive membrane permeability for Ca^{2+}-ions.

The part of our interests was deal with the influence of passive permeability on RyR that was expressed in sensibility of Ca^{2+}-transport to caffeine. Caffeine is the activator of intermediate state Ca^{2+}-channel gate of RyR, and respectively, it is the fine indicator of RyR functioning. However and the caffeine can affect to the manifestation of passive permeability. Also the significant factor can be the variation of ion concentration Mg^{2+}, without which Ca^{2+}-ATPase SR are not activated.

The increase of passive permeability we tried to cause by building in FSR membranes the unsaturated fatty acids. The lowering of passive permeability we tried to cause by extraction of FFA from membranes FSR by means of albumin, cleaned from related materials.

The fractions of FSR, received by the fragmentation of TC, in the presence of caffeine, reduce the efficiency of Ca^{2+}-accumulating that was shown by pH metric method [8]. And fractions, received from longitudinal tubules, didn't under the caffeine influences, when caffeine suspension (5–10 mM) was added to the incubating medium. As it can be seen from Figure 5.3, the addition Ca^{2+} on incubation medium, contained Mg-ATP, Na-oxalate and FSR, leads to sharp activating of acidification rate at incubating medium. After a time the rate of acidification drops to datum level, that indicates the accumulation of all amount of additional calcium by the reticulum bubbles. The rate of acidification of medium in absentia of Ca^{2+} is determined by the ATP hydrolysis by nonspecific enzyme ATPase by the essentially ion increment H^+ and respectively, inorganic phosphate in transport process of Ca^{2+}, it is possible to calculate the effectiveness of transport process (the mean the value of Ca^{2+}/ATP). Then, the larger amounts of phosphate are accumulated at incubation medium when transport additional amount of Ca^{2+} occurred, that effectiveness quantity of transport Ca^{2+}, the value Ca^{2+}/ATP, become smaller.

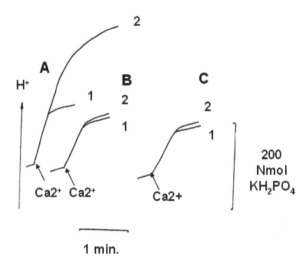

FIGURE 5.3   The caffeine influence to the Ca^{2+} transport by vesicular fragments of fractions of terminal cisterns (A) and longitudinal tubules (B,C), B—control FSR; C—FSR, treated by lipids, prepared from terminal cisterns rabbit skeletal muscle. 1—control; 2—+ 5mM caffeine.

The value Ca^{2+}/ATP, become are smaller or bigger in dependence of passive permeability changes. And the disagglutinating action caffeine (unconjugation action to Ca^{2+}-transport/ATP hydrolysis effectively) may be mediated not only

RyR activation, But the raise of passive permeability may take some part at this unconjugation action.

By this, we tested the caffeine actions to the Ca^{2+}-releasing from FSR under the low Ca^{2+} concentrations in registration medium (lowered 10^{-8} M) and without ATP. At this case the Ca^{2+} accumulating action of Ca^{2+}-pump can not be activated. The FSR was loaded by Ca-oxalate. Than the kinetic of Ca^{2+} release was registrated by pH-metric method with EGTA at medium. Under the binding of 1 Ca^{2+}-ion by EGTA, two H^+ are released from EGTA molecule (pH 7,0-7,1). The pH lowered. The graphics of depending of passive Ca^{2+}-releasing rate/Mg^{2+} concentration from FSR are presented: FSR—TC (Figure 5.4) and FSR—LT (Figure 5.1).

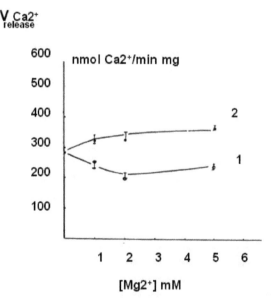

FIGURE 5.4 The action of Mg^{2+} to the caffeine influence to the Ca^{2+}-passive leaks from vesicular fragments of fractions of terminal cisterns. 1—control; 2—+ 5 mM caffeine.

The passive Ca^{2+}-releasing from FSR TC (Figure 5.4) is in bimodal depending of Mg^{2+} concentrations at medium. The variation of Mg^{2+} concentrations from 0 to 5 mM leads to lowering, and than to increasing of passive Ca^{2+}-releasing rate. The caffeine additions increased the passive Ca^{2+}-releasing rate, but the variation of Mg^{2+} concentrations operate weakly.

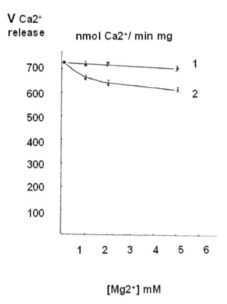

FIGURE 5.5   The action of Mg^{2+} to the caffeine influence to the Ca^{2+}-passive leaks from vesicular fragments of fractions of longitudinal tubules. 1—control; 2—+ 5 mM caffeine.

Data, presented at Figure 5.5 showed us that the passive permeability of FSR LT changed very negligible under the Mg^{2+} concentrations from 0 to 5 mM. The caffeine (5 mM) additions not changed the passive Ca^{2+}-releasing too.

Thus, caffeine influenced to the passive permeability at FSR TC. The rate of passive Ca^{2+}-releasing was increased to 20 percent. But, this value doesn't very essential for value of Ca^{2+}/ATP-lowering under the caffeine actions. The value of Ca^{2+}/ATP lowered by three times under the 5 mM caffeine additions to the FSR TC (Figure 5.3). Also the value of Ca^{2+}-transition by ATPase Ca^{2+}-transporting activity is 10 times as bigger than the passive permeability. Under different sets of conditions, caffeine doesn't influence to the passive permeability at FSR TC, and it lowered the value of Ca^{2+}/ATP as usually.

It is known that membrane of SR contains 0, 65 lipids mg by 1 protein mg. The phosphatidylcholine and phosphatidylethanolamine were the predominant phospholipids of SR. And SR contains FFA less 2 percent. [11], [12]. The lipid composition FSR TC and LT is similar [13]. We investigated the FSR TC and LT lipid compositions for our case. Data presented at Figure 5.6.

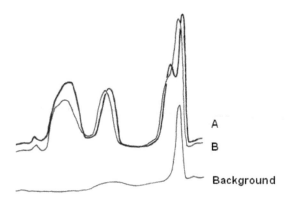

FIGURE 5.6   Scan of chromatographic plate with lipids, isolated from terminal cisterns (A) and longitudinal tubules (B), prepared from rabbit skeletal muscle. Chromatograms were carried out by methods [14], [15].
1 - Acidic phospholipids 6,3 percent (A); 5,0 percent (B)
2 - Phosphatidylcholine 43 percent (A); 35,6 percent (B);
3 - Phosphatidylethanolamine 22,1 percent (A); 25,2 percent (B)
4 - Free fatty acids (FFA) 19,4 percent (A); 20,1 percent (B)
5 - Neutral lipids 9,1 percent (A); 14,1 percent (B)...........

It is known that the FSR contain the 3–8 mg FFA/1 mg protein. The unsaturated FFA exists at those membranes [16]. As the membranes were washed from contaminated or integrated into bilayer FFA, the properties of passive permeability were changed. By these manipulations we used the FSR treatment with cleaned albumin.

The serum albumins are known as the carriers for any hydrophobic substances in blood. Respectively albumins also shall adsorb and meet to needed targets the biologically active substance of exogenous origin too. The level of saturation of albumin by hydrophobic molecules may be regulated. In our work the HSA was freed fully from adsorbed hydrophobic ligands if HSA was treated with the water suspension of pharmaceutical activated carbon when low the pH-values (4–5) applied, that allowed <to deploy> the molecule of HSA. Carbon was precipitated by centrifugation. The reverse of pH up to physiology (6,5–7,0) resulted in solution receipt of native HSA, able to connect with a number of hydrophobe molecules. The albumin used for stabilizing membrane preparations by the extracting from bilayer of the FFA. FFAs are the chaotropic agents, disturbing the crystalline structure of bilayer. HSA extracted of FFA from membranes of intracellular organelles faction of SR TC, obtained from white rabbit muscles (Figure 5.7).

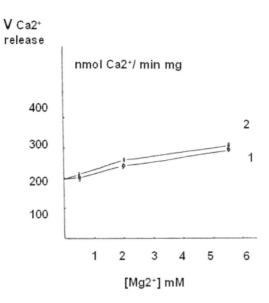

FIGURE 5.7 The action of Mg^{2+} to the caffeine influence to the Ca^{2+}-passive leaks from vesicular fragments of fractions of terminal cisterns. The free lipid acids were extracted from membranes of FSR with aid of clean albumin HSA. 1—control; 2— + 5mM caffeine.

The dependence of rate Ca^{2+}-releasing from ion concentration Mg^{2+} with and without caffeine from vesicles of fragmented terminal cisterns are submitted at Figure 5.7. The FSR TC membranes were washed from FFA.

In Table 5.1 are submitted the data, reflecting the dependence of value of Ca^{2+}/ATP from FFA presence at membranes of terminal cisterns when different ion concentrations Mg^{2+}. Data, presented at tables alludes to the fact that the presence or the absence of FFA at membranes of terminal cisterns in practice didn't influenced as on caffeine effect to the Ca^{2+}-transport, and on the Ca^{2+}-transport dependence from ion concentration Mg^{2+}. But this treatment increased the value of Ca^{2+}/ATP.

TABLE 5.1 The dependence of value of Ca^{2+}/ATP on the free fatty acid presence at terminal cisterns membrane under the two Mg^{2+} ions concentrations

Prepared FSR	1 mM Mg^{2+}		4 mM Mg^{2+}	
	–	5 mM caffeine	–	5 mM caffeine
FSR TC cleaned[a]	0,8+_0,05	0,3+_0,05	1,8+_0,05	0,8+_0,05
FSR TC	0,7+_0,05	0,25+_0,05	1,2+_0,05	0,35+_0,05

So that becomes clear that extraction of FFA from membranes of terminal cisterns in practice does not contribute to Ca^{2+}-ATPase function. Thus, the increase of passive membrane permeability for ions Ca^{2+} do not give distinguished contribution to tentatively observed the effectiveness of lowering of Ca^{2+}-transport under action of caffeine.

We indicated (using the pH-metric method) that in result, treatment by cleaned HSA is the decreasing of the passive permeability SR for Ca^{2+} that leads to increasing of the work efficiency of Ca^{2+} pump ATPase SERCA2.

The next experiments were the membranes incubation of FSR with free unsaturated fatty acids. This treatment considerable extent increases the caffeine effect on quantity of passive membrane permeability. It should be noted that the strengthening of influence of caffeine on the passive permeability by fatty acids was polymodal depends on ion concentration Mg^{2+} at registration medium. Maximum of caffeine effect on the passive the permeability occurs when 2мM Mg^{2+} (Figure 5.8). The dependence of rate of passive release Ca^{2+} from vesicles FSR TC from ion concentration Mg^{2+} with and without caffeine was shown at Figure 5.8. The FSR TC membranes were enriched by the free fatty acid.

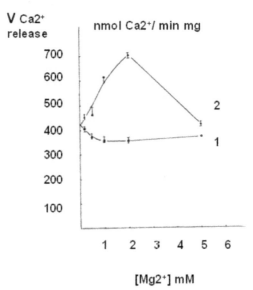

FIGURE 5.8 The action of Mg^{2+} to the caffeine influence to the Ca^{2+}-passive leaks from vesicular fragments of fractions of terminal cisterns. The free lipid acids were enriched the membranes of FSR. 1—control; 2—+ 5mM caffeine.

As it can be seen from Figure 5.8, when incubation (3hrs 10°C) the vesicles of FSR with linoleic acid (20 γ by 1 mg protein) was occurred, the passive releasing of Ca^{2+} from FSR increased. However in the presence of caffeine the rate of ion passive releasing Ca^{2+} becomes in depending on ion concentration Mg^{2+} in high degree. The curve had the extremum (maximum) in middle region of used concentrations Mg^{2+}. When ion concentration of Mg^{2+} increased from 0 up to 0, 5 MM, the caffeine in practice does not exert any effect at a rate of ion Ca^{2+} releasing by ion passive permeability under the presence of fatty acid at membranes. However at concentration of 2MM Mg^{2+} the rate of passive permeability by ion passive releasing Ca^{2+} increased in practice in twice. When the following increasing ion concentration Mg^{2+} occurred up to 5 MM, the reduction of caffeine effect on the passive permeability to ions Ca^{2+} by ion passive releasing $Ca^{2+}$was seen.

When we compared the data, presented at Figures 5.7 and 5.8, we may support that the passive ion Ca^{2+} releasing from vesicles of terminal cisterns decreases when extraction of FFA exists. The caffeine effect on the passive permeability to ions Ca^{2+} by ion passive releasing Ca^{2+} from terminal cisterns membranes, treated by the clean albumin for extraction of fatty acids, was reduced in practice up to 0.

5.4 CONCLUSION

The HSA was used for stabilizing membrane preparations by the cleaning of membrane from FFA, being the chaotropic agents that disturbed the crystalline structure of bilayer.

Aqua solution of HSA extracted FFA from membranes of intracellular organelles—two factions of fragmented sarcoplasmic reticulum (FSR). FSR were prepared of heavy and light fractions that were isolated from white rabbit muscles. It is known that membranes of SR contain 0,65 mg lipids by 1 mg protein. Of them, the FFA containing is less 2 percent. We obtain that pH-metering method was indicated that in result of HSA treating the several following processes existed: (1) the passive permeability FSR for Ca^{2+}, decreased; (2) the Ca^{2+}-pump ATPase (SERCA-2) conjugation increased; (3) the ion yield of transport Ca^{2+} through the Ca^{2+}-channel of Ryanodine receptor intensified. Similar extraction FFA by albumins from membranes can come about and in the animal's body.

The extraction of FFA from biological membranes in the animal body cells may lead to the more successful works of integrate enzymes (ATP-ase, for example), or ion channels. At firs case, the ion pumping occurred more effectively. At second case, the regulation processes for channel's activity may be mediated by ions streams through membrane. Thus the quantity reducing of ions Ca^{2+} in the environment leads to increasing of the work efficiency of Ryanodine receptor. Ca^{2+} ion releasing through the Ca^{2+} channel of Ryanodine receptor increased.

We have to note that the Mg^{2+}-ions concentration varied from 0,1 mM up to 10 mM was the main regulator factor for the passive permeability FSR for Ca^{2+}. But this problem may be discussed at next works.

We used caffeine in our work, as the test of RyR function activation. RyR activation was reflected in lowering of accumulation efficiency of calcium ions to the SR Ca^{2+}-store. The application of biologically active substances, which regulating the work of intracellular organelles, is good step for exploration functioning and components interconnection of these organelles. The active and passive Ca^{2+}-transitions in/out the SR-store and its crosstalk are the famous problem of interrelationships between RyR and Ca^{2+}-ATPase pump activations. These transitions are regulated by the many factors. Thus RyR activation was initiated by endogenous factor cADPR [17]. At the other following studies was found that RyR activation by cADPR was a result of Ca^{2+}ATPase-pump activation, mediated by increased luminal calcium. The Ca^{2+}-content at SR store and Ca^{2+}-releasing were regulated by luminal Ca^{2+}-sensitive leak [18]. These processes are actual, as for mammals, and for insects. Because there are the similarities of insect and mammalian ryanodine binding sites [19]. Thus the investigations of activators and inhibitors machineries mechanisms of RyR, Ca^{2+}-ATPase or Ca^{2+}-nonspecific leaks play the great role for the muscle and cardiovascular problems solutions. And it can help to found the target sites for insecticides.

Recently it was being found novel exogenous activator of RyR—flubendiamide. This substance stabilizes the insect RyRs in an open state in a species-specific manner and desensitizes the calcium dependence of channel activity [20]. Flubendiamide stimulated the Ca^{2+}-pump activity by decreasing in luminal calcium, which may induce calcium dissociation from the luminal Ca^{2+}-binding site on the Ca^{2+}-pump. This mechanism, as the authors [20] believe, should play an essential role in precise control of intracellular Ca^{2+}-homeostasis.

The investigations of each next activator or blocker actions permit us to understand the SR component interaction better, for potential application these substances as pharmacological substances in medicine, so and as insecticides in agriculture.

KEYWORDS

- **Human serum albumin**
- **passive permeability**
- **sarcoplasmic reticulum**

REFERENCES

1. Alekseeva, O. M.; and Kim, Yu. A.; The Influence of Caffeine Analogues and Antagonists on the Ca^{2+}-Accumulation by Sarcoplas Mic Reticulum at Skeletal Muscle, G. E. Zaikov (Ed.). Nova Science Publishers: New York, **2008**; pp 120–125.

2. Vekshina, O.; (Alekseeva), Kim, Yu.; Vekshin, N.; "Magic" calcium gradient for the operation of the sarcoplasmic reticulum". In Progress in Biochemical Physics, Kinetics and Thermodinamics. G. E. Zaikov (Ed.) Nova Science Publishers: New York, **2008**; pp 141–155.

3. Ikemoto N.; and Yamamoto, T.; "The luminal Ca^{2+} transient controls Ca^{2+} release/re-uptake of sarcoplasmic reticulum. *Biochem. Biophys. Res. Commun.* **2000**, *279*, 858–863.

4. Ritov, V. B.; Budina, N. B.; and Vekshina, O. M.; (Alekseeva) "The action of caffeine and Mg^{2+} on the efficacy of Ca^{2+} transport by terminal cisterns and longitudinal tubules of rabbit muscle". *Bull. Exp. Biologii i medizini.* **1985**, 1, 53–54.

5. Alekseeva, O. M.; and Ritov, V. B.; Two forms of Ca-dependent ATPase of sarcoplasmic reticulum. *Biochimia.* **1979**, *44*, 1582–1593.

6. Ikemoto N.; Kim, D. H.; and Antoniu, B.; Measurement of calcium release in isolated membrane of sarcoplasmic reticulum". *Method. Enzymol.* **1988**, *157*, 469–480.

7. Alekseeva, O. M.; Kim, Yu. A.; Rykov, V. A.; Vekshin, N. L.; The Quenching of Intrinsical Fluorescence of Sarcoplasmic Reticulum for the Lipid-Protein Interrelationship Determination. Handbook of Chemistry, Biochemistry and Biology: New Frontiers Nova Science Publishers: New York, Ed. G. E. Zaikov and A. N. Goloschapov. **2009**; pp 130–138.

8. Vekshin N. L.; Photonics of biopolimers. Springer. Biological and Medical Physics Series; **2002**.

9. Ritov V. B.; Acetylcholine influence and caffeine on functional activity fragmented the sarcoplasmic reticulum. *Biochimia*, **1971**, *36*, 393–399.

10. Folch, J., Lees, M., and Stanley, G. H. S.; A simple method for the isolation and purification of total lipids from animal tissue". *J. Biol. Chem.* **1957**, *226*, 497–509.

11. Meissner, G.; and Flaisher, S.; Characterization of sarcoplasmic reticulum from skeletal muscle. *BBA*, **1971**, *241*, 356.

12. Sarzala, M. G.; and Pilarska, Z. E.; Drabikowski, M.; Solubilization and reaggregation of SR membranes". *Protid. Biol. Fluid. Oxford.* **1974**, 109–113.

13. Lau, H.; Caswell, A. H.; Brunschwig, J-P; Baerwald, R. J.; and Careia, M.; Lipid analysis and freeze-fracture studies on isolated transverse tubules and sarcoplasmic reticulum subfractions of skeletal muscle". *J. Biol. Chem.* **1979**. *254*(2), 540–546.

14. Wagner, H.; Borhammer, Z.; and Wolf, P.; Dunnschichtromatographie von Phosphatiden und Glykolipiden. *Biochem.Zeitschr.* **1961**, *334*(2), 1175–1184.

15. Amenta, J. S.; A rapid chemical method for quantification of lipids separated by thin-layer chromatography". *J. Lipid. Res.* **1964**, *5*, 270–272.

16. Sarzalam, M. G.; and Drabikowski, W.; Free fatty acids as a factor modifying properties of fragmented SR during aging. *Life Sci.* **1969**, 8(2), 477–483.

17. Lukyanenko, V.; Gyöˮrke, I.; Wiesner, T. F.; and Gyöˮrke, S.; Potentiation of Ca^{2+} release by cADP-ribose in the heart is mediated by enhanced SR Ca^{2+} uptake into the sarcoplasmic reticulum. *Circ. Res.* **2001**, *89*, 614–622.

18. Lukyanenko, V.; Viatchenko-Karpinski, S.; Smimov, A.; Wiesner, T. F.; and Gyöˮrke, S.; Dynamic regulation of sarcoplasmic reticulum Ca^{2+} content and release by luminal Ca^{2+}-sensitive leak in rat ventricular myocytes. *Biophys. J.* **2001**. *81*, 785–798.

19. Lehmberg, E.; and Casida, J. E.; Similarity of insect and mammalian ryanodine binding sites. *Pestic. Biochem. Physiol.* **1994**, *48*, 145–152.

20. Masaki, T.; Yasokawa, N.; Tohnishi, M.; Nishimatsu, T.; Tsubata, K.; Inoue, K.; Motoba, K.; and Hirooka, T.; Flubendiamide, a novel Ca^{2+} channel modulator, reveals evidence for functional cooperation between Ca^{2+} Pumps and Ca^{2+} release. *Mol. Pharmacol.* **2006**, *69*(5), 1733–1739.

CHAPTER 6

A RESEARCH NOTE ON FEATURES OF WATER VAPOR SORPTION OF CHITOSAN MEDICINAL FILMS

ANGELA S. SHURSHINA[1], ELENA I. KULISH[1], and
SERGEI S. KOLESOV[2]

[1]Bashkir State Universite, Russia, Republic of Bashkortostan, Ufa, 450074, ul. Zaki Validi, 32

[2]The Institute of Organic Chemistry of the Ufa Scientific Centre the Russian Academy of Science, Russia, Republic of Bashkortostan, Ufa, 450054, October Prospect 71

CONTENTS

6.1   INTRODUCTION

F8The number of papers devoted to studying of processes of diffusion of elec-
trolytes in the polymer, including the diffusion of water, is traditionally great.
It is bound to that diffusion plays a prime role in such processes as a dialysis, a
permeability biological membranes and fabrics, prediction of protective properties of polymeric coverings, etc. [1]. The studying of process of diffusion of
water in a polymer matrix is necessary for the creation of medical polymer films
with controlled release of medicinal preparation, as a necessary condition providing diffusion transport of the drug from the polymer matrix, it is a swelling in
water. Under the influence of molecules of water diffusing in the film polymer
matrix can undergo various changes associated with restructuring of a material.
These effects can carry both reversible (pore filling) and irreversible (hydrolysis) character. In turn, any changes in the structure of the polymer can affect on
the process of sorption and swelling. [2] Thus, the sorption method can be used
as a tool allowing to track changes in the polymer matrix [3] The aim of this
work was to study the regularities of water sorption in drug polymer films.

6.2   EXPERIMENTAL

The object of investigation was a chitosan (ChT) specimen produced by the
company "Bioprogress" (Russia) and obtained by acetic deacetylation of crab
chitin (degree of deacetylation ~84 %) with $M_{sd} = 33,4000$. As the medicinal
substance (MS) used an antibiotics—amikacin (AMS) and cefazolin (CFZ).
Chemical formulas of objects of research and their symbols used in the text are
given in Table 6.1.

TABLE 6.1   Formulas research objects and symbols used in the reaction schemes

Formula object of study	Symbol
	$\text{wwwNH}_3^+\text{CH}_3\text{COO}^-$
monomer unit of acetate of chitosan	

amikacin sulfate

CFZ–Na⁺

cefazolin

ChT films were obtained by means of casting of the polymer solution in 1 percent acetic acid onto the glass surface with the formation of chitosan acetate (ChTA). Aqueous antibiotic solution was added to the ChT solution immediately before films formation. The content of the medicinal preparation in the films was 0.01, 0.05, and 0.1 mol/mol ChT. The film thickness in all experiments was maintained constant and equal to 100 microns. For prevention of dissolution of a film in water, we carried out isothermal annealing film samples at temperature 120°C.

Studying of interaction MS with ChT was carried out by the methods of IR- and UV-spectroscopy. IR-spectrums of samples wrote down on spectrometer "Shimadzu" (the tablets KBr, films) in the field of 700–3,600 cm⁻¹. UV-spectrums of all samples removed in quartz ditches thick of 1 cm concerning water on spectrophotometer "Specord M-40" in the field of 220–350 nanometers.

With the aim of determining the amount of medicinal preparation held by the polymer matrix β there was carried out the synthesis of adducts of the ChT-antibiotic interaction in acetic acid solution. The synthesized adducts were isolated by double repreciptation of the reaction solution in NaOH solution with the following washing of precipitated complex residue with isopropyl alcohol. Then the residue was dried in vacuum up to constant mass. The amount of preparation strongly held by chitosan matrix was determined according to the data of the element analysis on the analyzer EUKOEA—3000 and UF—spectrophotometrically.

The relative amount of water m_t absorbed by a film sample of ChT, determined by an exiccator method, maintaining film samples in vapors of water before saturation, and calculated on a formula: $m_t = (\Delta m)/m_0$, where m_0—the initial mass of ChT in a film, Δm—weight the absorbed film of water by the time of t time.

Structure of a surface of films estimated by the method of laser scanning microscopy on device LSM-5-Exciter (Carl Zeiss, Germany).

6.3   DISCUSSION

The main mechanisms in water transport in polymer films are simple diffusion and the relaxation phenomena in swelling polymer. In this case, the kinetics of swelling of a film is described by the equation [4]

$$mt/m_\infty = ktn \qquad (6.1),$$

where m_∞—relative amount of water in equilibrium swelling film sample, k—a constant connected with parameters of interaction polymer—diffuse substance, n—an indicator characterizing the mechanism of transfer of substance. If the transport of substance is carried out by classical diffusion mechanism and submits to the Fick's law, the index of n should to be close to 0.5. If transfer of substance is limited by the relaxation phenomena in polymers, for sorption kinetics characterized by a very slow final stage of establishment of equilibrium and values of n >0.5.

The parameter n determined for a film of pure ChT is equal 0,63 (i.e., >0,5) that is characteristic for the polymers, being lower than vitrification temperature [5] (Table 6.2). It is connected with slowness of relaxation processes in glassy polymers. If the film sample of ChT subjected to isothermal annealing, which is accompanied by relaxation of nonequilibrium conformations macrochains, values of n decreases and becomes close to the value of n = 0.5. The latter indicates that water transport is limited by diffusion, which in these conditions submits to Fick's mechanism.

Except a relaxation, process of isothermal annealing is accompanied by chemical reaction of sewing together of macrochains [6–8]. Similar chemical modification of the polymer matrix is accompanied by a change in the supramolecular structure of ChT in the film and creates some difficulties on paths of a stream of a diffusant. These difficulties, in turn, cause a decrease of an index of n to values, smaller, than 0.5.

The role of the polymer matrix modifier can play molecule of MS. Apparently from Table 6.2 data, the more MS entered into the film, the it is less value of an index of n and consequently the greater the deviation of the diffusion of the classical mechanism of Fick.

TABLE 6.2   Parameters of swelling of chitosan films in water vapor

composition of the film	The concentration of MS in the film, mol / mol ChT	Annealing time, min.	m_y, g/g ChT	n
ChT	0	15	2.50	0.50
		30	2.48	0.48
		60	2.47	0.44
		120	2.46	0.43
ChT-AMS	0.01	30	1.58	0.34
	0.05	30	1.36	0.30
	0.10	30	1.07	0.27
ChT-CFZ	0.01	30	2.14	0.44
	0.05	30	1.25	0.42
	0.10	30	0.94	0.40

All features inherent anomalous (not fikovsky) diffusion, described well by the relaxation model [9]. Unlike the Fickian diffusion assuming the instantaneous establishment and change of the surface concentration of sorbate, the relaxational model assumes change of concentration in the surface layer (i.e., variable boundary conditions) on the equation of first order:

$$c = c_0 + (c_\infty - c_0)(1 - \exp[-t/\tau]) \qquad (6.2)+$$

where c_0—the surface concentration of diffusant at $t \to 0$, c_∞—equilibrium concentration of the diffusing substance in the polymer, τ—the constant of proportionality characterizing average time of a relaxation of macromolecules. The physical sense of the Eq. (6.2) is that it allows to consider influence on a kinetics of a sorption of at the same time flowing past processes of structural relaxation. Thus the variable of the limiting sorption acts as the characteristic of a condition of the supermolecular structure of a material which reflects reorganizations in structure of samples, stimulated the absorbed sorbent. Anomalous diffusion with low n is characterized by a rapid decline of the boundary condition (6.2) [9]. Usually this occurs when the rate of relaxation processes V_R in the surface layer of the sorbent is higher or comparable to the rate of diffusion V_D molecules of sorbate in the polymer [10]. A similar anomaly was observed in [11] for the system polyvinylpyrolidone—water. For other glassy polymers is usually realized ratio $V_R < V_D$. A possible cause of the observed phenomena can be physical or chemical interaction of MS with the polymer matrix. For example, it is possible to present that in a polymeric matrix molecules of MS withheld a ChT physically "fill" porous surface structure in film material and create difficulties on paths of a stream of a diffusing. As a result, the rate of water absorption of the polymer matrix at large times of experiment will be significantly lower than in the initial time.

In Figure 6.1 presents profiles of a surface of films of ChT and its mixtures with MS at which comparison clearly visible that during the annealing process and, especially, at addition of a medicinal preparation, there is a change of a relief of a surface that is obviously bound as to flowing past processes of a relaxation, and "filling" of a pors.

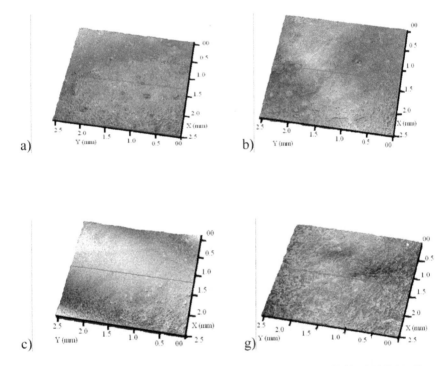

FIGURE 6.1 Micrograph of the surface (in contact with air) film individual ChT (a), films of ChT subjected to isothermal annealing within 1 hr (b), and films of ChT-AMS (c) and ChT-CFZ (g).

For example, an actual surface area of a film of initial ChT owing to a roughness and porosity of a surface in 4.8 times more theoretical, annealed ChT—by four times. In case of films ChT -MS differences in values of the total surface area of a film less and make 2.9 and 3.1 times for system ChT-AMS and ChT-CFZ, respectively. Note that the anomalously low values of n were observed also by other researchers, explaining them with the strong interaction of MS with polymer [12].

The structure of the used drugs allows to assume their chemical interaction with ChT, for example, formation complexes of the ChT—antibiotic by hydrogen bonds and/or the polymeric salts formed as a result of an exchange interaction. About interaction between antibiotics and ChT indicate data of UV-spectroscopy. The maximum of absorption of CFZ at a concentration 10^{-5} mol/l in 1 percent acetic acid is observed at 273 nm. At addition in solution of he equivalent quantity of ChT, intensity of peak of absorption of a medicinal preparation considerably increases, and the absorption of maximum bathochromic moves approximately on 5–10 nanometers. UV-spectrum of the AMS at a concentration of 10^{-2} mol/l in 1 percent acetic acid is characterized by a maximum of absorption at 267 nm. Addition to a solution of the AMS solution of ChT leads to precipitation, however the analysis of the supernatant shows that in its spectrum there is a shift of a maximum of absorption of the corresponding strip on 5–7 nanometers. The observed changes clearly indicate to impact of ChT on the electronic system of MS and the formation of adducts. Binding energies of the complexes, estimated by the shift of the absorption of maxima in the UV-spectra, make about 10 kJ/mol. It allows to assume that the complexing is carried out by means of hydrogen bridges. Interaction of studied antibiotics with ChT is confirmed by IR-spectroscopy data. For example, in the IR-spectrum of ChT absorption bands in the field of 1,640 and 1,560 cm^{-1} corresponding to the bending vibrations of acetamido and amino groups and the presence of bands at 1,458 and 1,210 cm^{-1} planar deformation vibrations of the hydroxyl groups. The analysis of IR-spectrums of adducts of interaction of ChT with antibiotics allows to speak about a number of the changes happening in IR-spectrums. So, in IR-spectrums of the ChT-CFZ polymeric adducts there are absorption bands 1,750 cm^{-1} and 1,710 cm^{-1}, corresponding to the stretching vibrations of $C = O$ and $C = N$ groups of medicinal compounds, and in IR-spectrum of the reaction product ChT-AMS band appears absorption of high intensity 619 cm^{-1} for the group SO_4^{2-}. Besides, in IR-spectrums of all analyzed compounds the absorption band broadening in the range of 3,000-3,500 cm^{-1}, corresponding to stretching vibrations of OH and NH-groups, in comparison with the corresponding strips of antibiotics and ChT is observed that in aggregate allows to tell about formation of complex compounds the ChT-antibiotic by means of hydrogen bridges.

Exchange interactions are not less essential to interpretation of data on diffusion between ChT and MS, especially in case of AMS. Exchange interactions between ChT and AMS may occur under the scheme:

Owing to two-basicity sulfate anions, it is possible to assume formation of two types of the salts providing sewing together of macromolecules of ChT with loss of its solubility. First, the water-insoluble "double" salt—sulfate ChT-AM, and secondly, the salt mixture - insoluble sulfate of ChT and soluble acetate of AM.

In case of CFZ, the exchange reaction between acetate of ChT and CFZ will be reduced to formation of soluble salts. Respectively, the reaction product ChT-MS in this case will consist of the H-complex ChT-CFZ.

Data on a share of the antibiotic connected in polymeric adducts (β), received in solutions of an acetic acid, are presented in Table 6.3.

TABLE 6.3   Mass fraction of the antibiotic β, defined in reaction adducts obtained from 1 percent acetic acid

Used antibiotic	The concentration of MS in the film, mol/mol ChT	β
AM	1.00	0.72
	0.10	0.33
	0.05	0.21
	0.01	0.07
CFZ	1.00	0.14
	0.10	0.08
	0.05	0.04
	0.01	0.02

As can be seen from Table 6.3, due to the fact that AM is able to "sew" chitosan chain is significantly more closely associated with macromolecules MS than for CFZ.

ACKNOWLEDGMENTS

Thus, the interaction ChT with MS actually leads to structural changes in the polymer matrix. This modification can be either chemical ("sewing" chitosan chains by antibiotic AMS) or physical (adsorption on the porous surface). Obviously, the modification is not only superficial but also a volume character. Changes in structure, in turn, cause difficulties to the flow of penetrant and cause deviations regularities of water sorption of chitosan films from classic fikovsky mechanism. Moreover, there is quite a reasonable correlation between the processes of water sorption medicinal films and processes of transport of molecules MS from this films [5, 13, 14]. Considering that from the methodical point of view process of absorption by a film of water is much simpler, than studying of kinetics of an exit of a medicinal preparation from a film, the getter method can be successfully used for research and prediction of effects of prolongation of action of medicinal films.

KEYWORDS

- **Chitosan**
- **Medicinal substance**
- **Sorption**

REFERENCES

1. Iordanskii, A. L.; Shterenzon, A. L.; Moiseev, Yu. B.; and Zaikov, G. E.; *Uspehi himii.* **1979**, XLVIII, №.8, 1461–1491.
2. Zaikov, G. E.; Iordanskii, A. L.; Markin, V. S.; *Diffuziya electrolitov v polimerah*, M.: Himiya, **1984**, 240 p.
3. A. Ya.; Malkin, and Chalyih, A. E.; *Diffuziya i vyazkost polimerov. Metodyi izmereniya,* M.: Himiya, **1979**, 304 p.
4. Hall, P. J.; Thomas, K. M.; and Marsh, H.; *Fuel.* **1992**, *71*, 11, 1271.
5. Chalyih, A. E.; *Diffuziya v polimernyih sistemah*, M.: Himiya, **1987**, p. 136.
6. Zotkin, M. A.; Vihoreva, G. A.; and Ageev E. P. et al. *Himicheskaya tehnologiya*, **2004**, *9, 15.*
7. Ageev, E. P.; Vihoreva, G. A.; and Zotkin M. A.; et. al. *Vyisokomolekulyarnyie soedineniya,* *46*, **2004**, *12*, 2035.
8. Smotrina, T. V.; *Butlerovskie soobscheniya*, *29*, 2, 98–101, **2012**.

9. Pomerancev, A. L.; Metodyi NelineinogoRregressionnogo Analiza dlya Modelirovaniya Kinetiki Himicheskih I Fizicheskih Processov, **2003**, Ph. D. Thesis, MGU.
10. Crank, J.; The Mathematics of Diffusion. Oxford: Clarendon Press, **1975**; 414 p.
11. Chalyih, A. E.; Gerasimov, V. K.; and Scherbina A. A. et al. *Vyisokomolekulyarnyie soedineniya*. **2008** *50*(6), 977–988.
12. Singh, B.; and Chauhan, N.; *Acta Biomaterialia*. **2008,** *4*, 1, 1244.
13. Zhulkina, A. L.; Ivancova, E. L.; and Filatova A. G.; et. al. *Kristallografiya*. **2009**, *54*(3), 497–500.
14. Ivancova, E. L.; Kosenko, R.Yu.; and Iordanskii A. L. et al. *Vyisokomolekulyarnyie soedineniya*. **2012**, *54,* 2, 215–223.

CHAPTER 7

A RESEARCH NOTE ON CREATION OF FILM CHITOSAN COVERINGS WITH THE INCLUDED MEDICINAL SUBSTANCES

A. S. SHURSHINA and E. I. KULISH

Bashkir State University Russia, Republic of Bashkortostan, Ufa, 450074, ul. Zaki Validi, 32; E-mail: alenakulish@rambler.ru

CONTENTS

Medicinal film coatings, on the basis of chitosan and antibiotics both of cephalosporin series and of aminoglycoside, have been considered. It has been shown that the antibiotics released from a film will be determined by the amount of antibiotics connected with chitosan by hydrogen bonds, on the one hand, and by the state of the polymer matrix, on the other.

7.1   INTRODUCTION

Decrease in effectiveness of therapy by the antibiotics, being observed recently, is caused, generally distribution of strains of bacteria steady against them. Polymeric derivants of antibiotics can help with the solution of this task [1]. Advantages of use of polymeric derivative antibiotics are obvious in that case when polymer carrier of medicinal substance is in a soluble form. However, it is not less important to consider that case when polymer carries out a matrix role—the carrier of medicinal substance. In this work some approaches to creation of sheet antibacterial coverings on a basis chitosan (ChT) of the prolonged action suitable for treatment of surgical, burn and slow wounds of a various etiology are considered. The choice as the ChT carrier is not casual as this polymer possesses the whole range of the unique properties doing it by irreplaceable polymer [2] for medicine.

7.2   EXPERIMENTALS

The objects of investigation were a ChT specimen produced by the company "Bioprogress" (Russia) and obtained by acetic deacetylation of crab chitin and antibiotics both of cephalosporin series—cephazolin sodium salt (CPhZ), cephotoxim sodium salt (CPhT), and of aminoglycoside series—amikacin sulfate (AMS), gentamicin sulfate (GMS). The investigation of the interaction of medicinal preparations with ChT was carried out according to the techniques described in [2, 3].

ChT films were obtained by means of casting of the polymer solution in acetic acid onto the glass surface with the formation of chitosan acetate (ChTA). The polymer mass concentration in the initial solution was 2 g/dl. The acetic acid concentration in the solution was 1, 10 and 70 g/dl. Aqueous antibiotic solution was added to the ChT solution immediately before films formation. The content of the medicinal preparation in the films was 0.1 mol/mol ChT. The film thickness in all the experiments was maintained constant and equal to 0.1 mm. The kinetics of antibiotics release from ChT film specimens into aqueous medium was studied spectrophotometrically at the wave length corresponding to the maximum absorption of the medicinal preparation.

In order to regulate the ChT ability to be dissolved in water the anion nature was varied during obtaining ChT salt forms. So, a ChT-CPhZ film is completely soluble in water. The addition of aqueous sodium sulfate solution, in the amount of 0.2 mol/mol ChT to the ChT-CPhZ solution, makes it possible to obtain an insoluble ChT-CPhZ-Na$_2$SO$_4$ film. On the contrary, a ChT-AMS film being formed at the components ratio used in the process of work isn't soluble in water. Obtaining a water-soluble film is possible if amikacin sulfate is transformed into amikacin chloride (AMCh). In this case the obtained ChT-AMCh film will be completely soluble in water. Thus, the following film specimens have been analyzed in the investigation: ChT-CPhZ and ChT-CPhT (soluble forms); ChT-CPhZ-Na$_2$SO$_4$ (insoluble-in-water form); ChT-AMCh (soluble form); ChT-AMS and ChT-GMS (insoluble-in-water forms).

With the aim of determining the amount of medicinal preparation held by the polymer matrix there was carried out the synthesis of adducts of the ChT-antibiotic interaction in the mole ratio 1:1 in acetic acid solution. The synthesized adducts were isolated by double reprecipitation of the reaction solution in NaOH solution with the following washing of precipitated complex residue with isopropyl alcohol. Then the residue was dried in vacuum up to constant mass. The amount of preparation strongly held by chitosan matrix was determined according to the data of the element analysis on the analyzer EUKOEA—3000.

7.3   THE RESULTS DISCUSSION

On the basis of the chemical structure of the studied medicinal compounds [4], one can suggest that they are able to combine with ChT forming polymer adducts of two types—ChT-antibiotics complexes and polymer salts produced due to exchange interaction. As a result some quantity of medicinal substance will be held in the polymer chain. The interaction taking place between the studied medicinal compounds and ChT was demonstrated by UV- and IR-spectroscopy data. The interaction energies evaluated by the shift in UV-spectra are about 7–12 kj/mole, which allows us to speak about the formation of complex ChT-antibiotic compounds by means of hydrogen bonds.

Table 7.1 gives the data on the amount of antibiotics determined in polymer adducts obtained from acetic acid solution.

TABLE 7.1   The amount of antibiotics determined in reaction adducts

C_{CH_3COOH}, g/dl in the Initial Solution	The Antibiotics Used	The Amount of Antibiotics in Reaction Adduct, % mass.
1	CPhZ	10,1
	CPhT	15,9
	AMS	61,5
	GMS	59,4
10	CPhZ	5,88
	CPhT	57,5
	AMS	55,8
	GMS	31,3
70	CPhZ	3,03
	CPhT	3,7
	AMS	41,3
	GMS	40,1

Attention should be paid to the fact that the amount of medicinal preparation in the adduct of the ChT-medicinal preparation reaction is considerably higher in the case of antibiotics of aminoglycoside series than in the case of antibiotics of cephalosporin series. This can be connected with the fact that CPhZ and CPhT anions interact with ChT polycation forming salts readily soluble in water. In the case of using AM and GM sulfates because of two-base character of sulfuric acid one may anticipate the formation of water-insoluble "double" salts—ChT-AM or ChT-GM sulfates due to which additional quantity of antibiotics is held on the polymer chain.

Table 7.2 gives the data on the value of the rate of AM and GM release from film specimens formed from acetic acid solutions of different concentrations. The rate was evaluated only for water-insoluble films because at using soluble films the antibiotic release was determined not by medicinal preparation diffusion from swollen matrix but by film dissolving.

TABLE 7.2   Transport properties of chitosan films in relation to medicinal preparation release

Acetic Acid Concentration g/dl	The Antibiotics Used	Release, % mass./h for Chitosan Specimens
1	AMS	0,5
	GMS	0,4
10	AMS	0,8
	GMS	0,5
70	AMS	1,5
	GMS	1,3

Attention must be given to the fact of interaction between the rate of anti-biotics release from chitosan films and their amount which is strongly held in ChT chain. For example, at increasing the concentration of acetic acid used as a solvent the amount of medicinal preparation connected with the polymer chain decreases in all the cases considered by us. Correspondingly, the rate of antibiotics release from films insoluble in water, increases.

The influence of the amount of medicinal preparation strongly held in ChT matrix, on the rate of medicinal substance release from the film must be most pronounced at comparing the rates of release of antibiotics of aminoglycoside series and cephalosporin one. However, ChT-CPhZ and ChT-CPhT films are soluble in water while ChT-AMS and ChT-GMS ones do not dissolve in water and it isn't correct to compare them. At ChT transition into insoluble form (by adding sodium sulfate) the rate of release of antibiotics of cephalosporin series decreases considerably (Figure 7.1, curve 1) as compared with a soluble form but still it is higher than that in the case of antibiotics of aminoglycoside series (Figure 7.1, curve 2). It should be also noted that the rate of antibiotics release from soluble ChT-CPhZ film (Figure 7.1, curve 3) is also higher than in the case of ChT-AMCh film (Figure 7.1, curve 4). Thus, considerable differences between the rate of release of aminoglycoside series antibiotics and that of cephalosporin series antibiotics is evidently explained by the difference in the amount of ChT-antibiotics adduct.

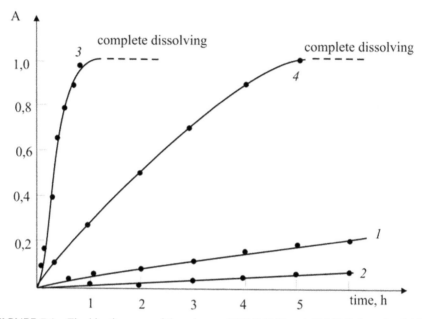

FIGURE 7.1 The kinetic curve of the release of CPhZ (1,3) and AM (2,4) from insoluble (1,2) and soluble (3,4) films.

Thus, at forming film coatings one should proceed from the fact that a medicinal preparation can be distributed in the polymer matrix in two ways. One part of it connected with polymer chain, for example, by complex formation is rather strongly held in that polymer chain. The rest of it is concentrated in polymer free volume (in polymer pores). The rate of release of antibiotics from the film will be determined by the amount of antibiotics connected with ChT by hydrogen bonds, on the one hand, and by the state of the polymer matrix including its ability to dissolve in water, on the other hand.

    The work has been carried out due to the financial support of the RFFR and the republic of Bashkortostan (grant_r_povolzhye_a № 11-03-97016)

KEYWORDS

- **Chitosan**
- **Medicinal preparation**
- **Modification**
- **State of polymer matrix**

REFERENCES

1. Skryabin, K. G.; Vikhoreva, G. A.; and Varlamov, V. P.; Chitin and chitosan. Obtaining, properties and application. M.: Nauka, **2002**, 365 p.
2. Mudarisova, R.Kh.; Kulish, E. I.; Kolesov, S. V.; and Monakov, Yu. B.; Investigation of chitosan interaction with cephazolin. *JACh.* **2009**, *82* (5), 347–349.
3. Mudarisova, R.Kh.; Kulish, E. I.; Ershova, N. R.; Kolesov, S. V.; and Monakov Yu. B.; The study of complex formation of chitosan with antibiotics amicacin and hentamicin. *JACh.* **2010**, *83*, *6*, 1006–1008.
4. Mashkovsky M. D.; Medicinal preparations. Kharkov: Torsing, **1997**, *2*, 278 p.

PART II
BIOLOGICAL MATERIALS

CHAPTER 8

DIMEBON EFFECT ON THE FLUIDITY OF MICE SYNAPTOSOMAL MEMBRANE BY EPR SPIN LABELING METHOD

N. YU. GERASIMOV*, O. V. NEVROVA, V. V. KASPAROV, A. L. KOVARSKIJ, A. N. GOLOSHCHAPOV, and E. B. BURLAKOVA

Emanuel Institute of Biochemical Physics RAS, 119334, Kosygina str., 4, Moscow, Russia;

*E-mail: n.yu.gerasimov@gmail.com

CONTENTS

8.1   INTRODUCTION

In the last few years antihistamine drug Dimebon is proposed for the treatment of neurodegenerative disorders such as Alzheimer's disease. In [1] it was shown that the drug at low concentrations is AMPA-kainate receptors protogonist and NMDA-receptors antagonist, whose action mechanism remains unknown. It is assumed that Dimebon exhibits neuroprotective properties and improves cognitive function in dementia [2–4].

The lipids composition and lipid bilayer structure of membrane significantly affects on the proteins functional activity [5]. Earlier we showed disorders in the membranes structural characteristics as a result of the Alzheimer's disease development [6]. In addition, Burlakova E. B. proposed membrane memory model [7], in which the determining factor is the structure of the membrane, and fluidity is one of the important structural characteristics of lipid bilayer. Therefore it is important to study the neuroprotectors action on the structural state of the lipid bilayer, accordingly in the work was investigated the Dimebon effect on the fluidity of mice synaptosomal membrane by EPR spin labeling method.

8.2   EXPERIMENTAL

Dimebon was kindly provided by Bachurin S.O., IPAC RAS (Figure 8.1). The drug was injected abdominally every day in the concentration 1mg/kg. Samples were taken in 1, 3, 7, and 15 days after injection. Female of HLK white outbread mice 20–23 g in size were used as experimental animals.

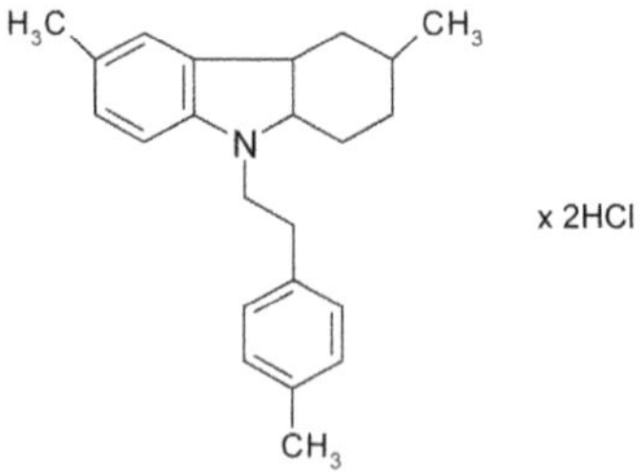

FIGURE 8.1   Dimebon.

The sample was a synaptosomes combined fraction isolated from the brain of 6–8 animals. Each measerment was carried out 4–5 times. Synaptosomes were separated by differential centrifugation in sucrose [8].

The fluidity of the membranes lipid bilayer was determined by the method of electron paramagnetic resonance (EPR) of spin probes. As probes it were used the stable nitroxide radicals 2,2,6,6-tetramethyl-4-capryloyl-oxypiperidine-1-oxyl (probe I) and 5,6-Benzo-2,2,6,6-tetramethyl-1,2,3,4-tetrahydro- γ -carboline-3-oxyl probe II), synthesized at the Institute of chemical physics named after N.N. Semenov RAS (Figure 8.2).

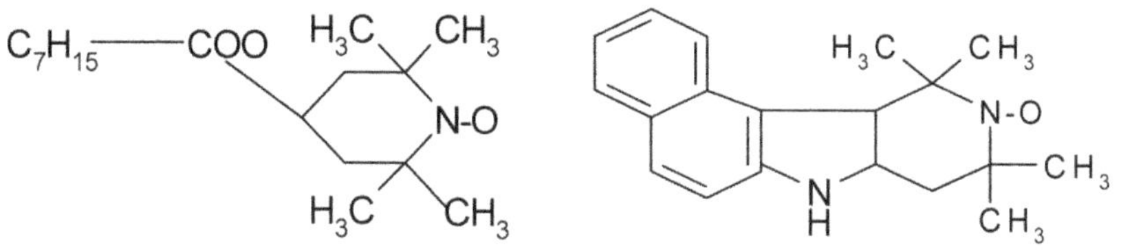

Probe I   Probe II

FIGURE 8.2   Spin probes.

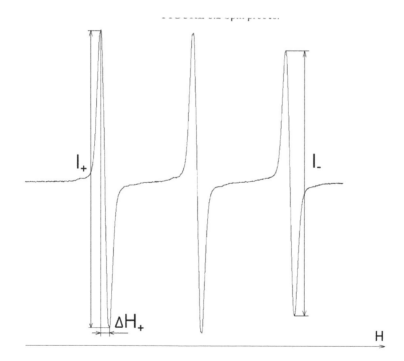

FIGURE 8.3   Typical ESR spectrum for probe I and II.

In [9], it is shown that the probe I is localized mainly in the surface layer of membrane lipid components, and probe II—in lipids, close to the proteins, that allow to talk about the lipid-protein interactions in the membranes based on the probes I and II behavior in the lipid bilayer. For convenience, we will call probe I as "lipid", and probe II as "protein" subsequently.

The rotational diffusion correlation time (τ_c), characterizing the membrane components microviscosity, was calculated from EPR spectra (Figure 8.3) by formula

$$\tau_c = 6.65 \times 10^{-10} \times \Delta H_+ \times ((I_+/I_-)^{0.5}-1),$$

presented in [10]. The EPR spectra were registered in the temperature range of 283–317 K (10-44°C) on radiospectrometer ER 200D-SRC "Brucker" in the magnetic spectroscopy center of the Emanuel Institute of Biochemical Physics RAS.

The well-known relation by Stokes-Einstein (see, e.g., [11]) binds the parameter τ_c with medium viscosity, surrounding the probe, by formula $\tau_c = \eta V/kT$, where V—volume of the radical (it can be considered directly proportional to the molecular weight); η- dynamic viscosity of the medium; k - Boltzmann constant and T—absolute temperature. Dynamic viscosity η is related to temperature by the following empirical relation $\eta = A'e^{b/T}$ [12], whence it follows $\ln\tau_c = A'' + b/T + \ln(1/T)$, where A', A", b—constants. Investigated in our work temperature range (from 283 to 317 K) is sufficient narrow insomuch, that component $\ln(1/T)$ changes is very low in comparison with the term b/T, so we can assume $\ln\tau_c = a + b/T$.

Thus, the experimental curves should be linearized in the coordinates $\ln\tau_c$ and 1/T. However, like such behavior is taken place only for simple, one-component system. Membrane structures are systems, characterizing by the presence of thermoinducible structural transitions. Accordingly, dependency plot of $\ln\tau_c$ from 1/T for such structures must be a polyline, which break points are the points of structural transitions [13]. The slope coefficient of these lines allows to determine the transition activation energy $\Delta E_a = bR$ [14], where b—slope coefficient of the corresponding straight-line section, and R—absolute gas constant. Activation energy corresponds to the transition energy of 1 mole of membrane lipids [14].

Statistical data processing was carried out by the methods of parametric statistics using software Microsoft® Office Excel and Origin® 6.1 with the statistical reliability 95 percent.

8.3 RESULT AND DISCUSSION

In this paper, we studied the effect of neuroprotector Dimebon on synaptosomal membrane microviscosity isolated from the mice brain in 1, 3, 7, and 15 days after chronic injection of drug. An example of the rotational diffusion correlation time dependence on temperature is shown in Figures 8.4 and 8.5, in coordinates ln (τ) and 1/T. As it seen from figures, the dependence is a polyline, consisting of inclined and practically horizontal sections. Horizontal sections correspond to generalized structural changes of corresponding membranes areas (near protein—for probe II, Figure 8.5; and lipid—for probe I, Figure 8.4). Two transitions are observed in both areas of membranes at temperatures 289–293 K (16-20°C) and 311–317 K (38–44°C) for the control group. The first transition is associated with alterations in the lipid phase, and the second with the changes of the proteins structure [15, 16]. On the other hand, after chronic injection of Dimebon an additional structural transition is appeared in synaptosomal membranes near protein areas in the temperatures interval 297–301 K (24–28°C). This fact apparently is associated with the integration of the present substance in the protein structures, including receptors and channels [2], that leads to changes of membrane proteins structure and their lipid environment.

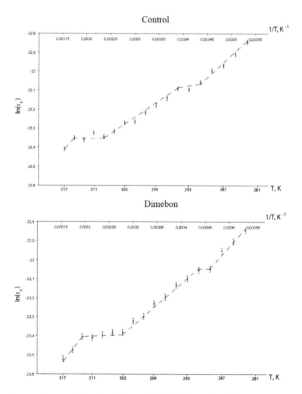

FIGURE 8.4   Dependence of the ln (t_c) on 1/T of the probe I in synaptosomal membranes after 7 days of the Dimebon daily injections.

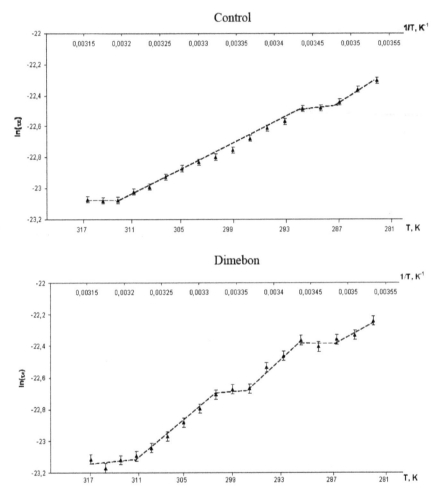

FIGURE 8.5   Dependence of the ln (t_c) on 1/T of the probe II in synaptosomal membranes after 7 days of the Dimebon daily injections.

The lipid-protein free phase of membranes are remained almost untouched (wasn't differ from the control). Possibly, Dimebon cannot get out easily from protein. Arising in this case strong bonds of Dimebon with proteins leads to the appearance of additional structural transition in the membranes [2].

Table 8.1 shows the corresponding structural transitions for all periods of drug injunction with the activation energies. The table demonstrates that significant structural changes take place only at long-term delivering of agents compared with the control. For short-term drug injunction (1-3 days), the substance

TABLE 8.1 Structural transitions in the synaptosomal membranes under Dimebon daily injection with their activation energy. (*Statistical validity < 90%)

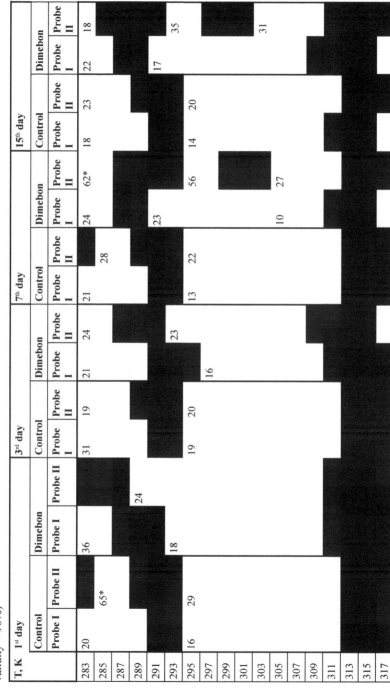

T, K	1st day				3rd day				7th day				15th day			
	Control		Dimebon		Control		Dimebon		Control		Dimebon		Control		Dimebon	
	Probe I	Probe II	Probe I	Probe II	Probe I	Probe II	Probe I	Probe II	Probe I	Probe II	Probe I	Probe II	Probe I	Probe II	Probe I	Probe II
283	20	65*	36		31	19	21	24	21	28	24	62*	18	23	22	18
285				24												
287																
289																
291			18		19	20	16	23	13	22	23	56	14	20	17	35
293	16	29														
295																
297																31
299																
301																
303											10	27				
305																
307																
309																
311																
313																
315																
317																

is excreted from the body and not accumulating in organism, without significant influence. Neuroprotector concentration is leveled off apparently at a constant value after the long periods of injection, whereby the effect from the drug action are increased. The activation energies of the corresponding transitions are not significantly changed and stay within the measurement accuracy close to the control, except the cases with additional transition. Additional transition appearance denotes a significant change of the membranes near protein areas structure, that indicate about changes of the protein structure and their lipid environment

The effect of the Dimebon on the synaptosomal membrane fluidity is shown in Figure 8.6. Values of the rotational diffusion correlation time at a temperature of 297 K, at which there is no thermo-induced structural changes in membranes, were taken as an indicator of microviscosity. The figure shows dynamic changes in near protein areas microviscosity with time, while these changes are minor in lipid areas.

The dependence of the relative changes of rotational diffusion correlation times from the agent introduction time at 297 K is shown on Figure 8.7. The fluidity of membrane near protein areas are distinctly increased immediately after the first injection of Dimebon, that is, a response to external action (Figures 8.6 and 8.7). Fluidity is returned to the level of control at a further injection. It is assumed that after a single injection organism is trying to exclude an foreign agent, and at the same time membranes structure is changed. An organism cannot removes the drug fully due to the strong bonds formation of Dimebon with proteins, that leads to a change in the protein structure, and, as consequence, lipid environment of protein changes its structure, in such a way that the structural characteristics are returned to norm. Subsequently, the body stops to responding on Dimebon injection, leading to an additional phase transition in the region 297–301 K (Figure 8.5, Table 8.1). In our opinion such influence can lead to appearance of adverse effects with long-term use of the drug. Apparently, a certain level of fluidity is important for cells, because membranes lability was being returned to normal (Figures 8.6 and 8.7) with time.

Thus, we can assume that the membranes microviscosity is an important structural characteristics of membranes, and plays an important role in cell metabolism.

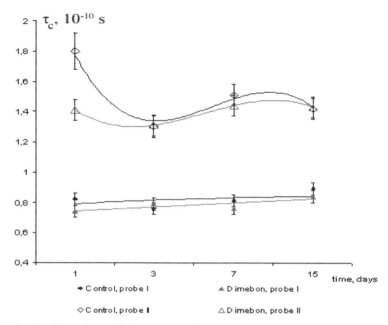

FIGURE 8.6 Dependence of the t_c on time of the Dimebon daily injection for synaptosomal membranes at the temperature 297 K.

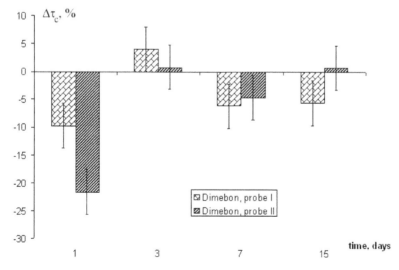

FIGURE 8.7 Dependence of the t_c changes relative to control on time of the Dimebon daily injection for synaptosomal membranes at the temperature 297 K.

CONCLUSIONS

We have found that, neuroprotector Dimebon changes the membranes structure after chronic injection in such a way that the lipid bilayer fluidity is returned to normal with time. Therefore, membranes microviscosity plays an important role in the cell metabolism and is an important structural characteristic. Thus, at the drugs selection for the diseases therapy, it should to take into account changes of the lability and membranes structure that probably will allow to avoid of adverse effects and to raise the effectiveness of medicinal product.

KEYWORDS

- **Dimebon**
- **Lipid-protein interactions**
- **Membranes fluidity**
- **Membranes structure**
- **Spin probe**

AIM AND BACKGROUND

Membrane structure play important role in the development of dementia [6]. Therefore, it was important to investigate the membrane structural changes under neuroprotector injection.

REFERENCES

1. Grigorev, V. V.; Dranyi, O. A.; and Bachurin, S. O.; Comparative study of action mechanisms of dimebon and memantine on AMPA- and NMDA-subtypes glutamate receptors in rat cerebral neurons. *Bull. Exp. Biol. Med.* **2003**, *136*(5), 474–477.
2. Bachurin, S.; Tkachenko, S.; Baskin, I.; Lermontova, N.; Mukhina, T.; Petrova, L.; Ustinov, A.; Proshin, A.; Grigoriev, V.; Lukoyanov, N.; Palyulin, V.; and Zefirov, N.; Neuroprotective and cognition-enhancing properties of MK-801 flexible analogs. Structure-activity relationships. *Ann N Y Acad Sci.* **2001**, *939*, 219–236.
3. Grigoriev, V. V.; Proshin, A. N.; Kinzirskii, A. S.; and Bachurin, S. O.; Binary mechanism of action of cognition enhancer NT1505 on glutamate receptors. *Bull. Exp. Biol. Med.* **2012**. *153*(3), 298-300.
4. Cano-Cuenca, N.; Solis-Garcia Del Pozo, J. E.; and Jordan, J.; Evidence for the efficacy of latrepirdine (dimebon) treatment for improvement of cognitive function: a meta-analysis. *J. Alzheimers. Dis.* **2013**, epub.
5. Robert, B.; Gennis: Biomembranes. Molecular Structure and Functions. Springer-Verlag; **1997**

6. Gerasimov, N.Yu.; Goloshhapov, A. N.; and Burlakova, E. B.; Structural state of erythrocyte membranes from human with Alzheimer's disease. Khimicheskaja fizika, **2009,** *28*,7, 82–86 (in Russian)

7. Burlakova, E. B.; The role of the membrane lipids in the process of the information transfer. Zhurn. fiz. khimii. **1989,** *18,* 1311 (in Russian)

8. Prohorova, M. I.; Metody biohimicheskih issledovanij. Izd-vo Leningrad. un-ta, **1982,** (in Russian)

9. Binjukov V. I.; Borunova S. F.; and Gol'dfel'd M. G.; i dr. Study of the structural transitions in the biological membranes using the method of spin probes. *Biokhimija.* **1971** *36*(6), 1149, (in Russian)

10. Vasserman A. M.; Buchachenko A. L.; Kovarskij, A. L.; and Nejman, I. B.; Study of the molecular motion in the fluids and polymers using the method of the paramagnetic probe . *Vysokomolekuljarnye soedinenija.* **1968** *10A,* 1930 (in Russian)

11. Kuznecov, A. N.; *Metod spinovogo zonda.* M. Nauka, **1976** (in Russian)

12. Kuhling, H.; Spravochnik po fizike. M.: Mir, **1983,** (in Russian)

13. Chapman, D.; Phase transitions and fluidity characteristics of lipids and cell membranes. *Quart. Rev. Biophys.* **1975,** *8.* 85–191.

14. Shinitzky M.; and Inbar, M.; Microviscosity parameters and protein mobility in biological membranes. *Biochimica et Biophysica Acta.* **1976,** 133–149.

15. Gendel' L. Ja.; Gol'dfel'd M. G.; Kol'tover V. K.; Rozancev Je. G.; and Suskina, V. I.; Investigation of he conformation transitions in the biological membranes using the method of the weakly bound spin probe. *Biofizika.* **1968,** 13(6), 1114–1116 (in Russian)

16. Goloshhapov, A. N.; and Burlakova, E. B.; Thermo induced structural transitions in the membranes after antioxidants injection and with malignant grow. *Biofizika.* **1980,** *25*(1), 97–101 (in Russian)

CHAPTER 9

BIOCONVERSION OF SOLID ORGANIC WASTES BY MAGGOTS OF BLACK SOLDIER FLIES (*Hermetia illucens*)

A. I. BASTRAKOV, A. A. ZAGORINSKY, A. A. KOZLOVA, and N. A. USHAKOVA*

A. N. Severtsov Institute of Ecology and Evolution, RAS, 33 Leninskij prosp., Moscow, 119071, RUSSIA, Fax: (8495) 954-55-34, *E-mail: naushakova@gmail.com

CONTENTS

9.1   INTRODUCTION

Invertebrates get the increasing attention due to accumulation of a huge mass of organic waste and rising the scarcity of feed protein. Special attention should be paid to the organisms which are able to grow on a variety of organic substrates, both of animal and plant origin. Invertebrates, which are appropriate for commercial breeding, include earthworms, flies (*Hermetia illucens, Musca domestica, Calliphora* sp., *Sarcophaga* sp.), Tenebrionidae beetles (*Tenebrio molitor, Zophobas morio*), Gryllidae (*Acheta domesticus, Gryllus bimaculatus*), etc. Insects' protein is considered as alternative to fish meal's protein, fats contain bioactive unsaturated fatty acids, and chitin of cuticle has immunomodulatory properties.

Hermetia illucens is a large fly of Stratiomyidae family. In the last decade *Hermetia illucens* gained widespread popularity in the world, because its maggots can recycle various organic substrates such as pig and bird manures, food, and agricultural waste. Biomass produces by maggots is often used as part of ration of agricultural animals: pigs, birds, fish. In order to study the decomposition process of the substrate it is necessary to analyze the dynamics of substrate utilization, accumulation of larvae's biomass, also to identify relative indicators of substrate conversion and biochemical composition of maggots [1, 3, 8]. Scientists from many countries of the world are engaged in breeding of *Hermetia illucens*. Some literary sources describe the possibility of adding larvas to animals' ration. From 1980, he U.S. pays considerable attention to problems of introducing of maggots in feed for pigs [7] and fish [2]. Canada has conducted a big research in studying the decomposition process of organic waste by *Hermetia illucens* larvae in the North [1]. Nowadays Sweden studies the hygienic aspect of larvae's using in feed of farm animals and their waste as biofertilizers [5]. During the observation of larvaes growing in chicken litter [4] it was noted that *E. coli* and *Salmonella enterica* were being significantly suppressed due to the presence of antimicrobial proteins in insects [6] and the presence of symbiotic bacterias which produce antibiotics, including *Lysobacter*. *Lysobacter* was found in adults, prepupae and pupae. This fact explains the potential antimicrobial effect of Black soldier flies when colonizing substrates [9].

The aim of the study is to investigate the process of bioconversion of plant substrates by *Hermetia illucens*'maggots.

9.2   MATERIAL AND METHODS

9.2.1   BLACK SOLDIER FLIES

Hermetia illucens passes five stages of the life cycle: egg, larva, prepupa, pupa and adult. Lifecycle of *Hermetia illucens* from egg to egg takes an average of

45 days. Larval period of Hermetia illucens is 20–25 days. Lifecycle scheme of Hermetia illucens is shown in Figure 9.1.

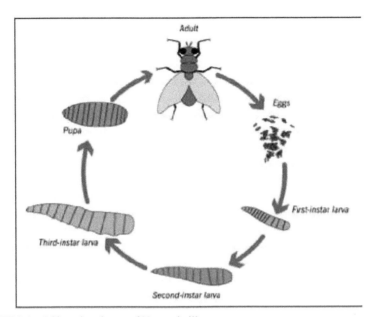

FIGURE 9.1   Lifecycle scheme of Hermetia illucens.

Flies mating and oviposition take place in liquor. The eggs are incubated. After larvaes' appearance, the latter are placed in bioreactor, where the bioconversion process is held. When larvae pass into the stage of prepupal, they are separated and dried. The remaining mass of recycled substrate and exogenous secretion of larvae are composted.

9.2.2   EXPERIMENT

Hermetia illucens were stored in Insectarium of Institute of ecology and evolution, RAS. Wheat bran and corn grits were used as feed substrates. Experimental samples together with 1–2 old age larvaes were placed in plastic containers. Stocking density of larvae is five individuals per 1 cm². The height of the substrate layer was 5 cm. Room temperature was maintained at 28°C. Each variant of the experiment was replicated three times. The bioconversion process was regarded as completed when the proportion of prepupae was at least 50 percent. Larvae were separated from the recycled substrate by sieving. Larvae and substrate were weighed on digital weigher with accuracy of 0.001gr and then were dried at temperature of 70°C.

Main indicators:

$$D = \frac{W - R}{W}$$

D is the substrate conversion
W is the initial dry weight of substrate
R is the final dry weight (the amount used)
Substrate consumption per day (gram per day):

$$WRI = \frac{D}{t} * 100$$

Weight gain of the maggots is the difference between the final and initial dry weight of larvae.

Daily weight gain of larvae (gram per day) is the ratio of the weight gain of the maggots to the conversion duration.

The feed consumption ratio is the ratio of the consumed feed (predetermined substrate minus final weight) to the increase in dry biomass of maggots (final biomass minus initial biomass).

Composition and nutritional value of fodder grown on them larvae were determined by standard methods.

9.3 RESULTS AND DISCUSSIONS

The main results of the process of bioconversion of wheat bran and corn grits are shown in Table 9.1. One maggot of *Hermetia illucens* consumes almost two times more of corn grits than the bran per day. Conversion of corn grits in a part of dry matter was D 90 percent, whereas the bran—D 64 percent. Table 9.2 presents data on the chemical composition of larvae grown on two plant substrates. Feed substrate has a significant effect on conversion rates as well as the chemical composition of the larvae. In compare with wheat bran there is a high concentration of starch in corn grits, which leads to the increase of fat in larvae. Unlike corn grits when using bran for breeding larvae there was marked an increase in calcium content in larvae. Bran was used as the example to show high speed and efficiency of conversion of waste with high fiber content by *Hermetia illucens* maggots. The collected biomass of larvae has the following composition: high content of protein (over 45%), fat (up to 45%), calcium

(0.8%), and phosphorus (about 0.8%). However, in order to get larger amount of larvae's biomass, it is more preferable to use more nutritious substrates. The high fat (about 30%) and protein (about 40%) content in dry larvaes of *Hermetia illucens* determines its high potential for using in combined feed for animals. Therefore, such a source of animal protein is an attractive alternative for feed production, which currently is depend on fishmeal.

KEYWORDS

- **Bioconversion**
- ***Hermetia illucens***
- **Maggot**
- **Waste with high fiber**

REFERENCES

1. Alvarez, L.; Electronic theses and dissertations. Paper 402; **2012**.
2. Bondari, k.; and Sheppard, D. C.; *Aquaculture.* **1987** *24*, 103–109.
3. Diener, S.; Zurbrugg, C.; and Tockner, K.; *Waste. Manag. Res.* **2009**, *27*, 603–610.
4. Erickson, M. C.; Islam, M.; Sheppard, C.; Liao, J.; and Doyle, M. P.; *J. Food. Protect.* **2004**, *67*, 685–690.
5. Lalander, C.; Diener, S.; Magri, M. E.; Zurbrügg, C.; Lindström, A.; Vinneras B: Sci ;Total Environ. **2013**, *458–460*, 312–318.
6. Nattori, S.; Nippon Rinsho. **1995**, *5*, 1297–1304 ().
7. Newton, G. L.; Booram, C. V.; Barker, R. W.; and Hale, O. M.; *J. Anim. Sci.* **1977**, *44*, 395–399.
8. Newton, G. L.; Sheppard, C.; Watson, W.; Burtle, G.; and Dove R.; University of Georgia, College of Agriculture and Environment Science, Deptartment of Animal and Dairy Sci. Annual Report; **2004**.
9. Zheng, L.; Crippen, T. L.; Singh, B.; Tarone, A. M.; Dowd, S.; YU, Z.; Wood, T. K.; Tomberlin, J. K.; *J Med Entomol.* **2013**, *50*(3), 647–58.

The work was supported by the Program of fundamental research of the Department of Biological Sciences of RAS "Biological resources of Russia", 2014

CHAPTER 10

CHONDROGENIC DIFFERENTIATION OF HUMAN ADIPOSE TISSUE DERIVED STROMAL CELLS

I. P. SAVCHENKOVA

All Russian State Research Institute of Experimental Veterinary Medicine of Ya.R. Kovalenko, 109428, Russia, Moscow, Ryazanskiy pr. 24/1 , E-mail: s-ip@mail.ru

CONTENTS

10.1   INTRODUCTION

In recent years, several groups have demonstrated that mesenchymal cells isolated from the stromal vascular fraction (SVF) of human subcutaneous adipose tissue (SAT) display multilineage developmental plasticity in vitro [1–7]. The presence of an alternative, biologically accessible source of multipotent cells is a pressing matter for regenerative medicine [8]. The availability of an unlimited cell source replacing human chondrocytes could be strongly beneficial for cell therapy, tissue engineering, in vitro drug screening, and development of new therapeutic options to enhance the regenerative capacity of human cartilage. Articular cartilage exhibits little intrinsic repair capacity, and new tissue engineering approaches are being developed to promote cartilage regeneration using cellular therapies.

The goal of this study was to estimate the chondrogenic potential of cellular populations isolated from SVF of human SAT in vitro.

10.2   METHODS

10.2.1   SUBJECTS

Liposuction aspirates from SAT sites were obtained from healthy slightly overweight human donors (male and female) with their permission under local anesthesia. The patients were undergone elective procedures in local plastic surgical offices of Beauty plaza.

10.2.2   CELL ISOLATION

Cells were isolated from adipose tissue (AT) using methods previously described [1] with some modification. Briefly, harvested tissue was thoroughly washed with PBS, minced, and submitted to enzymatic treatment with 0.075 collagenase of the type I (Gibco, Invitrogene, Life Technologies, United States) solution on the basis of DMEM (Gibco, Invitrogene, Life Technologies, United States) at 37°C for 30 min. Collagenase was neutralized with equivalent volume of the nutritive medium DMEM added by 10 percent FCS (HyClone, Perbio, Belgium) and centrifuged at 1,000 g for 5 min. The supernatant, containing mature adipocytes, was aspirated. The pellet was identified as the SVF. The obtained sediment was resuspended in DMEM and passed through a filter with pores 80 μm (for SVF) and then through 10 μm. The cells were resuspended by centrifugation with an acceleration of 200 g for 5 min and resuspended in nutrition medium. Medium for the ATSCs cultivation was DMEM with low glucose content (1 g/l) added by 10 percent FCS, a single solution of nonessential amino acids, and antibiotics (Gibco Invitrogene, Life Technologies, United States). The final concentration of streptomycin in medium was 100 μg/ml and the final concentration of penicillin was 100 U/ml. The medium was replaced 24 h after seeding. Cells were plated at a density of 1×10^6 in a 10-cm dish. Attached cells were cultivated to improve cell yield. Morphologic characteristics of the cells were studied visually in native preparations and Giemsa stained preparations.

10.2.3   FLOW CYTOMETRY

Cell surface Ags expression was assayed with flow cytometry in an Epics Elite Coulter cytometer. Cells removed from the substrate with 0.25 percent trypsin solution were washed and counted. Samples with $2 \cdot 10^5$ cells were incubated with primary Abs in PBS with 2 percent FCS for 45 min at 4°C in the dark. Primary mouse Abs in a 1:30 dilution against the following human Ags were used: CD13, CD29, CD31, CD34, CD44, CD45, CD49b,d, CD73, CD90, CD105, CD133, CD 166, HLA ABC, HLA DR-II, HLA DP, and HLA DQ (Becton Dickinson). PE-labeled anti-mouse IgG (Becton Dickinson) in 1:30 dilutions were applied as secondary Abs.

10.2.4   INDUCTION OF CHONDROGENIC DIFFERENTIATION

The composition of the chondrogenic medium was the following: high (4.5 g/l) glucose DMEM (Gibco, Invitrogene, Life Technologies, United States), 10 ng/ml TGF-β1 (BD, Germany), 100 nM dexamethasone (KRKA, Slovenia), 50 μg/ml ascorbate2-phosphate (Sigma, USA), ITS (6.25 μg/ml insulin, 6.25 μg/ml transferrin, 6.25 μg/ml selenium acid final concentrations), 5.33 μg/ml linoleic acid (Gibco, USA), and 1.25 mg/ml human serum albumin (Biomed, Russia).

For chondrogenic differentiation, cells were cultured in a pellet cell culture. Therefore $1 \cdot 10^6$ cells were centrifuged in a 15-ml polypropylene tube at 300 g for 5 min to form a pellet. Without disturbing the pellet, the cells were cultured for 4 weeks in 1.0 ml of complete chondrogenic differentiation medium. The medium was replaced every 4 days.

After the culture period, cryosections 5–6 μm were analyzed with hematoxylin-eosin staining [9]. The sections were fixed with ice-cold methanol for 5 min. For immunohistological analysis Abs against type II collagen (Chemicon International Inc.) were used. Samples were stained with a kit produced by Novocastra Laboratories Ltd. Company (Great Britain) and a substrate produced by BD Biosciences Pharmingen Company (USA). Pellets cultured in medium without inductors were used as a control.

10.2.5   RNA ISOLATION AND RT-PCR

The isolation of RNA from cells was performed using the kit SV Total RNA Isolation System (Promega, United States) according to the manufacturer's recommendations. Induced and uninduced cells were trypsinized and washed twice with PBS. The cell sediment was added by 175 μl of lysing solution, then by 350 μl of buffer solution for dilution (SV Dilution Buffer); mixed accurately; and incubated for 3 min at 70°C. The mixture was centrifuged; supernatant was transferred into a clean test tube and added by 200 μl of 95 percent ethanol. The solution was transferred into a column (Spin Column Assembly) and centrifuged at 16,000 g for 1 min. The column was washed with 600 μl of solution for washing (SV RNA Wash Solution) and 50 μl of incubation mixture containing DNase I were applied. After 15 min of incubation at room temperature, 200 μl of Stop Solution (SV DNase Stop Solution) were added, centrifuged at 16,000 g for 1 min, and washed twice with solution (SV RNA Wash Solution) as described above. RNA was eluted with 100 μl of nuclease-free water and stored at –20°C. Concentration of isolated RNA was determined by measuring optical density at the wavelength of 260 nm from formula 1 OD = 40μg RNA/ml.

The PCR-reaction, the following primers for markers were used: Collagene type II (Col) , forward 5'-TTTCCCAGGTCAAGATGGTC-3' and backward 5'-CTTCAGCACCTGTCTCACCA-3' and

Glyceraldehyde-3-phosphate dehydrogenase (GAPDH), forward 5'-GGGCTCCTTTTAACTCTGGT- 3' and backward 5'-TGGCAGGTTTTTC-TAGACGG-3' [10].

Glyceraldehyde-3-phosphate dehydrogenase was used as a control for assessing PCR efficiency.

Incubation of the reactional mixture was performed under the following conditions: preliminary denaturing at 94°C for 3 min, then 35 cycles according to the following scheme: (1) denaturing at 94°C for 20 s, (2) annealing of primers at 55°C for 20 s, and (3) elongation at 72°C for 40 s. After the end of the cycles the final elongation was performed at 72°C for 5 min.

10.3   RESULTS

Cellular population isolated from SVF of human SAT had fibroblastic morphology and high adhesive property. Cytofluorometric analysis of these cells revealed expression of surface Ags common for human BM MMSC. Thus, the cells were positive for the following Ags: CD13 (99.0%), CD29 (96.5%), CD44 (96.4%), CD49b (80.6%), CD49d (95.1%), CD73 (98.8%), CD90 (99.6%), CD105 (98.9%), CD166 (97.9%), and HLA ABC (99.2%). The cells were negative for hemopoietic stem cell markers CD34 (0%), CD133 (1.7%), endothelium cell marker CD31 (0.1%), for hematopoietic cell-associated marker CD45 (0.1%) and histocompatible locus Ags HLA DR (0.1%), HLA DP (1.0%), and HLA DQ (0%). The data are presented for five donors. Figure 10.1 shows the Ags expression on the surface of ATSCs of a spontaneous choice donor. Cells with the same phenotype were used in our experiments.

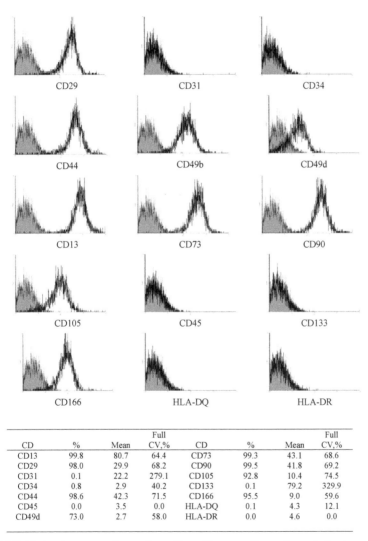

			Full				Full
CD	%	Mean	CV,%	CD	%	Mean	CV,%
CD13	99.8	80.7	64.4	CD73	99.3	43.1	68.6
CD29	98.0	29.9	68.2	CD90	99.5	41.8	69.2
CD31	0.1	22.2	279.1	CD105	92.8	10.4	74.5
CD34	0.8	2.9	40.2	CD133	0.1	79.2	329.9
CD44	98.6	42.3	71.5	CD166	95.5	9.0	59.6
CD45	0.0	3.5	0.0	HLA-DQ	0.1	4.3	12.1
CD49d	73.0	2.7	58.0	HLA-DR	0.0	4.6	0.0

FIGURE 10.1   Ags expression on the surface of human adipose tissue-derived stromal cells. (*Grey color*)—control staining with Ig labeled with phycoethythrin; (*white color*)—staining with specific Abs.

Figure 10.2 shows the results of the histological and immunohistological staining of the pellet cell culture samples. It is seen (Figure 10.2, b) that the cell morphology was typical for cells of articular cartilage. Most of the cells were round or oval shaped, with round or flat nuclei shifted to the edge of the cytoplasm. Some cells had cytoplasmic inclusions which appeared when the extracellular matrix secretion took place. In some cases, isogenic groups consisted of two cells were evident. We have chosen the type II collagen as specific marker to cartilage tissue. Immunohistochemical analysis of the sample sections on the 21th has revealed the type II collagen expression (Figure 10.2c).

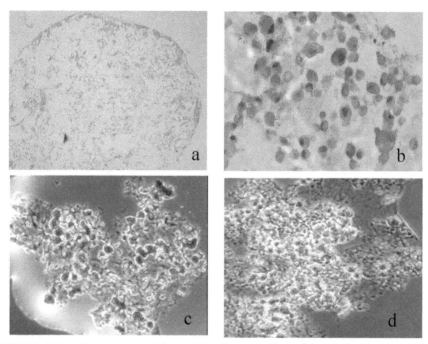

FIGURE 10.2   Chondrogenic differentiation of human adipose tissue-derived stromal cells. Pellets stained with hematoxylin-eosin (a, b) or with Abs against type II collagen (c)— culturing in the induction medium for 21 days. Pellets cultured in medium without inductor stained with Abs against collagen of type II—control (d). a.—Ob. 5x, oc.10x; b.—Ob.40x, oc.10x; c,d.—Ob.20x, oc.10x.

Analysis of gene expression in the process of differentiation of cells derived from SAT along the chondrogenesis has revealed that the cells isolated from SAT in control and in experimental groups were negative for type II collagen at the 14th day. In cells culturing without inductor, the product of type II collagen gene expression was not detected. The cartilage specific transcripts of type II collagen were detected in the cells submitted to differentiation at the 21th day (Figure 10.3).

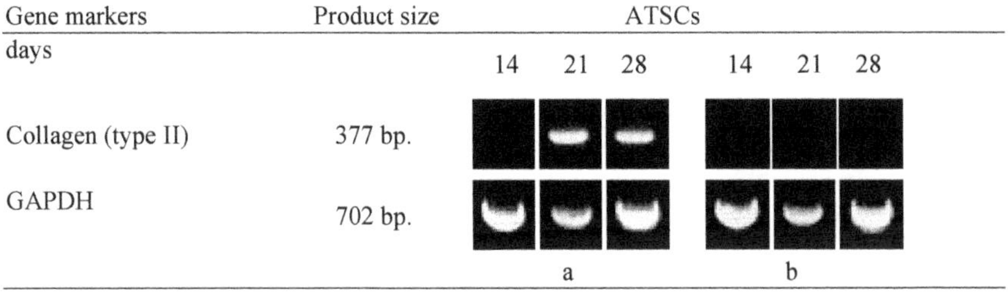

Gene markers	Product size	ATSCs					
days		14	21	28	14	21	28
Collagen (type II)	377 bp.						
GAPDH	702 bp.						
			a			b	

FIGURE 10.3   Real-time reverse transcription-polymerase chain reaction analysis of type II collagen expression in induced adipose tissue-derived stromal cells (a) and noninduced adipose tissue-derived stromal cells—control (b).

10.5   DISCUSSION

Previously, we isolated and characterized cells with phenotypes similar to multipotent mesenchymal stromal cells (MMSCs) derived from adult human bone marrow (BM) and SAT [11]. There are no known stem cell-specific markers for the identification of putative stem cells or progenitor cells within AT [12–15]. However, MMSCs derived from BM are considered to be positively stained by Abs, e.g., SH2, SH3, and SH4 [16]. An Ag binding SH2 Abs was identified as endoglin (CD105), the receptor for the transforming growth factor TGF- β1 [17]. SH3 and SH4 Abs recognize individual epitopes of the protein of 5′-nucleotidase (CD73), which is bound to the membrane and plays an essential role in the activation of B-lymphocytes. The surface Ag SB10 also is often used to identify human MMSCs [18]. The expression of this Ag is activated by molecules of leukocyte cellular adhesion ALCAM, CD166. In our experiments, cells obtained from SVF of human SAT yielded high expressions of these Ags. The cells were positive for α4 integrin (CD49d) and for molecule of cellular adhesion VCAM (CD106). Cellular populations isolated in our experiments were similarity with the stem cell population isolated from AT by deUgarte et al., 2003 [19].

In this study we have described the chondrogenic differentiation of human ATSCs in vitro. Previously we demonstrated the directed differentiation of MMSCs isolated from human BM in cells of cartilage tissue by culturing them in OPLA polymer three-dimensional scaffolds [20]. The induction of chondrogenesis in MMSC depends on many factors, such as cell density, cell adhesion, and growth factors [21–23]. High cell density, a serum-free medium, and specific growth factors are the necessary requirements for inducing chondrogenesis. For high cell density we used pellet cell culture and a serum-free medium added TGF–β1. In our experiments, the morphological assay of the pellet cell culture crysections indicated zones of cells morphologically similar to the cells of the cartilage tissue of the intermediate layer of the hyaline cartilage.

It is known, that extracellular matrix of the articular cartilage mostly consists of collagens (50–70%). The predominant collagen type in hyaline cartilage is collagen of type II, which comprises 95 percent of the cartilage. Therefore the type II collagen synthesis is considered to be a reliable marker of chondrogenic differentiation. In our experiments the type II collagen detected using immunohistochemistry and RT-PCR analysis. Immunostaining revealed the type II collagen in the extracellular matrix cells after maintenance in chondrogenic medium for 3 weeks. This fact confirmed present of product of type II collagen expression in ATSCs at the 21th days by RT-PCR analysis.

The sequential events in this pathway leading from the undifferentiated mesenchymal stem cells derived from BM to a mature chondrocytes were investigation by Barry et al. [24]. Comparative analysis of expression of type II collagen by MSCs grown in pellet culture in the presence of TGF-β1, -β2, or -β3 for 7, 14, and 21 days detected that the size of pellets and extent of type II collagen staining in the TGF-β2- and β3-treated cultures were greater than those treated with TGF-β1 at 14 and 21 days. Winter et al. [25] suggest that although the multilineage potential of bone marrow-derived stromal cells (BMSC) and ATSC was similar according to cell morphology and histology, some minor differences in marker gene expression occurred before and after induction of diverse differentiation pathways. Immunostaining of paraffin sections indicates that induction of type collagen expression occurs earlier in BMSC spheroids than in ATSC spheroids. Our data suggest that induction of type II collagen expression occurs later at 21 days in ATSC pellet culture. We not detected a transcript of type II collagen gene in the TGF-β1 treated culture at 14 days. These results suggest that it is necessary over a longer period then 2 weeks to promote type II collagen gene expression using these induction conditions. On the basis of our results, we propose that cellular populations isolated from SVF of SAT had potential for chondrogenic differentiation and should serve as candidate cells for tissue engineering applications.

KEYWORDS

- **Adipose tissue**
- **Chondrogenic differentiation**
- **Human, induction**
- **Multipotent mesenchymal stem cells**
- **Tissue engineering.**
- **Type II collagen expression**

REFERENCES

1. Zuk, P. A.; Zhu, M.; Mizuno, H.; Huang, J. I.; Futrell, W. J.; Katz, A. J.; Benhaim, P.; Lorenz, H. P.; and Hedric, M. H.; *Tissue. Eng.* **2001**, *7*, 211.
2. Halvorsen, Y. D.; Franklin, D.; Bond, A. L.; Hitt, D. C.; Auchter, C.; Boskey, A. L.; Paschalis, E. P.; Wilkison, W. O.; and Gimble J. M.; *Tissue. Eng.* **2001**, *7*, 729.
3. KATZ, J.; Methods of Tissue Engineering. Academic Press: NY, **2002**; 277–286.
4. Zuk, P. A.; Zhu. M.; Ashjian, P.; deugarte, D. A. Huang, J. I.; Mizuno, H.; Alfonso, Z. C.; Fraiser, J. K.; Benhaim, P.; and Hedrick, M. H.; *Mol. Biol. Cell*, **2002**, *13*, 4279.
5. Erickson, G. R.; Gimble, J. M.; Franklin, D. M.; Rice, H. E.; and Awad, H.; Guilak F.; *Biochem. Biophys. Res. Commun.* **2002**, *290*, 763.
6. Gimble, J. M.; and Guilak, F.; *Curr. Top. Dev. Biol.* **2003**, *58*, 137.
7. Aust, L.; Delvin, B.; Foster, S.; Halvorsen, Y.; Hicok, K.; Laney, Td. T.; SEN, A.; Willingmyre, G.; Gimble, J.; *Cytotherapy.* **2004**, *6*, 7.
8. Gimble, J. M.; Katz, A. J.; and Bunnell, B. A.; *Circ. Res.* **2007**, *100*, 1249.
9. Pierce, E. (Ed.); Histochemistry. Mir. Moscow, **1962**; 1031 p. (in Russian).
10. Caterson, E. J.; Nesti, L. J.; Danielson, K. G.; and Tuan, R. S.; *Mol. Biotechnol.* **2002**, *20*, 245.
11. Teplyashin, A. S.; Korjikova, S. V.; Sharifullina, S. Z.; Chupikova, N. I.; Rostovskaya, M. S.; Savchenkova, I. P.; *sitologiya.* **2005**, *47*(2), 130.
12. Katz, A. J.; Tholpady, A.; Tholpady, S. A.; Shang, H.; Ogle, R. C.; *Stem Cells.* **2005**, *23*, 412.
13. Festy, F.; Hoareau, L.; Bes-Houtmann, S.; Pequin, A-M.; Gonthier, M-P.; Munstun, A.; Hoarau, J. J.; Cesari, M.; and Roche, R.; *Histochem. Cell. Biol.* **2005**, *124*, 113.
14. MitchelL, J. B.; Mcintosh, K.; Zvonic, S.; Garrett, S.; Floyd, Z. E.; Kloster, A.; Halvorsen, Yu. D.; Storms, R. W.; Goh, B.; Kilroy, G.; Wu, X.; Gimble, J. M.; *Stem Cells.* **2006**, *24*, 376.
15. Astori, G.; Virgnati, F.; Bardelli, S.; Tubio, M.; Gola, M.; Albertini, V.; Bambi, F.; Scali, G.; Castelli, D.; Rasini, V.; Soldati, G.; Moccetti, T.; *J. Transl. Med.* **2007**, *5*(55), 10.
16. Haynesworth, S. E.; Baber, M. A.; and Caplan, A. I.; *Bone.* **1992**, *13*, 69.
17. Barry, F. P.; Boynton, R. E.; Haynesworth, S.; Murphy, J. M.; and Zaia, J.; *Biochem. Biophys. Res. Commun.* **1999**, *265*, 134.
18. Bruder, S. P.; Ricalton, N. S.; Boynton, R. E.; Connolly, T. J.; Jaiswal, N.; Zaia, J.; and Barry, F. P.; *J. Bone. Min. Res.* **1998**, *13*, 655.
19. Deugarte. D. A.; Alfonso, Z. C.; Zuk, P. A.; Elbarbary, A. ; Zhu, M.; Ashjian, P.; Benhaim, P.; Hedrick, M. H.; Fraser, J. K. *Immunol. Lett.* **2003**, 89, 267.
20. Teplyashin, A. S.; Korjikova, S. V.; Sharifullina, S. Z.; Rostovskaya, M. S.; Chupikova, N. I.; Vasyunina, N.Yu.; Andronova, N. V.; Treshalina, E. M.; Savchenkova, I. P.; *Cell. Tissue. Biol.* **2007**, *1*(2), 125.

21. Sekiya, I.; Vuoristo, J. T.; Larson, B. L.; and Prockop, D. J.; *PNAS.* **2002**, *99*, 4397.

22. Tuli, R.; Tuli, S.; Nandi, S.; Huang, X.; Manner, P. A.; Hozack, W. J.; Danielson, K. G.; Hall, D. J.; and Tuan, R. S.; *J. Biol. Chem.* **2003**, *278*, 41227.

23. Goessler, U. R.; Bugert, P.; Bieback, K.; Deml, M.; Sadick, H.; Hormann, K.; Riedel, F.; *Cell Mol. Biol. Lett.* **2005**, *10*, 345.

24. Barry, F.; Boynton, R. E.; Liu, B.; Murphy, J. M.; *Exp. Cell Res.* **2001**, *268*, 189.

25. Winter, A.; Breit, S.; Parsch, D.; Benz, K.; Steck, E.; Hauner, H.; Weber, R. M.; Ewerbeck, V.; and Richter, W.; *Arthritis & Rheum.* **2003**, *48*, 418.

CHAPTER 11

INVESTIGATION OF PROPERTIES OF FITO-STIMULATE CYANOBACTERIAL COMMUNITIES OBTAINED FROM LOWER VOLGA ECOSYSTEMS

BATAEVA YULIA

Federal State Budget Educational Institution of Higher Professional Education
Astrakhan State University, 414 025, Astrakhan, st. Tatishchev, Russia, 16 E, fl. 307;
E-mail: aveatab@mail.ru

CONTENTS

11.1   INTRODUCTION

Nowadays, agriculture is not the preferred chemical preparations, affecting negatively on soil fertility, environment, quality of products and biological agents that stimulate plant growth, yield environmentally friendly products and contribute significantly to soil fertility. These agents include the soil, rhizosphere, nitrogen-fixing bacteria that form numerous physiologically active substances that enter the roots of the plants and intensify the growth. They increase crop yields, reducing the ripening time, increase the nutritional value, improve resistance to disease, frost, drought and other adverse factors, speed up germination and rooting, reduce preharvest abscission of ovaries and abscission until the late frosts, are struggling with weeds and perform many other function.

A special place in the soil is occupied by cenoses algae and cyanobacteria. Cyanobacteria, unlike other soil algae fixed from the atmosphere, not only carbon but also atmospheric nitrogen, produce the biologically active substance and form of the primary production of organic matter [8, 9, 10]. Under natural conditions, cyanobacteria always develop in association with many other organisms, due to a mucous sheath, and, therefore, have excellent opportunities for adaptation and resistance to drastically changing physical and chemical environmental conditions. This creates the preconditions for more efficient appliances cyanobacterial communities (CBC) at their introduction in the soil. In addition, cyanobacteria are economical to cultivation and have high growth rates, which is essential for the production of biologics. With rapid growth rates of cyanobacteria accumulate 20 days to 15 tons of biomass per 1 hectare.

In agricultural biotechnology cyanobacteria are poorly understood, not counting the rice fields [6]. The possibility of using cyanobacteria as fertilizers extensively studied in Asian countries, mainly in rice fields. Soil nitrogen-fixing cyanobacteria living cultures have a positive effect on growth and yield of rice [3, 4, 7].

The objective of our work was to study the properties of laboratory phytostimulate CBC from various aquatic and soil ecosystems of the Astrakhan region on the seeds of cress.

11.2   RESULTS

To investigate the activityof phytostimulate used 25 laboratory collection of communities of cyanobacteria isolated from different aquatic and soil ecosystems of the Astrakhan region [1]. CLS collector supported by reseeding after 1–2 months on a liquid medium BG-11 in Erlenmeyer flasks 100–250 ml in volume and cultivation under natural light and a temperature of 22–25°C [8].

Cyanobacteria and algae are identified by morphological characters, using the determinant Gollerbah etc. [2], textbook Zenov, Shtina [5].

Study of phytotoxicity and phytostimulate activity was performed using the test on the seeds of cress. For the experiment, the toxicity of the seeds of cress were placed in humid rooms—sterile Petri dish with filter paper in triplicate. Presterilized seeds treated with 70 percent ethanol for 3–5 minutes, then washed 3 to 5 times with sterile distilled water. Each camera was placed 50 seeds, which are wetted with a slurry of 10 ml of sterile distilled water and 0.3 g biomass CBC pilot. The suspension was prepared by adding distilled water, chopped into small pieces of CBS strands (to reduce the gradient concentration of the suspension), then stirred for 3 min. Control seeds were soaked in sterile distilled water. Seeds treated with distilled water and suspensions were germinated for three days in daylight and 25°C.

Availability of growth-stimulate, inhibitory or neutral effect was determined by comparing the seed germination, root and stem length of plants in the control and experimental variants.

11.3 DISCUSSION OF RESULTS

In studying the structure and composition of the investigated CBC found a large variety of habitats cyanobacteria, green and found diatoms. The main share of uprepresentatives of cyanobacteria species of the genera: *Phormidium, Oscillatoria, Anabaena, Nostoc, Microcystis, Gloeocapsa* [1]. Fewer species are cyanobacteria genera *Chroococcus, Spirulina, Nostoc, Pleurocapsa, Synechococcus,* and *Synechocystis.* Among the green algae is commonly encountered genus *Chlorella, Chlorococcum.*

Results of the evaluation study of phytostimulate activity CBC presented in Table 1.

TABLE 11.1 Effect of bacterization of cyanobacterial communities in the germination of garden cress

Variant (№ Community of Cyanobacteria)	Seed Germination, %	The Average Size of the Root Length, mm	The Average Size of the Length of the Shoot, mm
Control	85,7 ± 1,2	25,7 ± 1,4	13,4 ± 0,6
1	90,0 ± 1,1	25,0 ± 1,2	13,1 ± 0,4
2	86,0 ± 1,1	31,9 ± 1,9	14,0 ± 1,0
3	86,0 ± 1,2	13,7 ± 0,6	10,8 ± 0,9
4	81,3 ± 1,7	26,3 ± 2,4	13,3 ± 1,1

TABLE 11.1   *(Continued)*

Variant (№ Community of Cyanobacteria)	Seed Germination, %	The Average Size of the Root Length, mm	The Average Size of the Length of the Shoot, mm
5	82,6 ± 2,4	31,4 ± 1,1	19,2 ± 0,2
6	87,3 ± 2,9	31,7 ± 1,2	13,0 ± 0,4
7	98,0 ± 1,5	28,7 ± 1,0	16,6 ± 0,5
8	88,6 ± 1,7	18,5 ± 0,2	11,2 ± 1,5
9	90,0 ± 1,1	19,4 ± 1,3	10,8 ± 0,7
10	86,0 ± 1,2	19,6 ±3,2	14,4 ± 1,4
11	83,6 ± 3,5	39,7 ± 1,8	15,4 ± 0,4
12	88,0 ± 1,1	31,0 ± 0,5	13,9 ± 0,7
13	86,6 ± 1,3	24,7 ± 3,5	13,7 ± 1,0
14	95,0 ± 0,6	38,0 ± 0,4	13,6 ± 0,5
15	78,6 ± 4,6	32,7 ± 1,5	13,8 ± 0,5
16	73,3 ± 0,6	26,7 ± 0,8	17,6 ± 0,8
17	80,0 ± 1,7	14,0 ± 0,5	13,0 ± 0,7
18	78,0 ± 2,5	13,9 ± 1,5	8,1 ± 0,6
19	82,0 ± 2,1	16,9 ± 0,5	11,4 ± 1,4
20	96,0 ± 1,4	23,6 ± 0,5	12,0 ± 1,5
21	84,0 ± 1,1	43,7 ± 0,2	17,5 ± 0,3
22	82,0 ± 1,2	15,9 ± 1,5	14,4 ± 1,6
23	82,0 ± 1,2	18,0 ± 1,2	14,4 ± 1,5
24	64,8 ± 3,1	4,1 ± 1,2	8,0 ± 0,6
25	84,0 ± 1,8	15,8 ± 1,8	12,1 ± 0,7

As a result, data processing, were investigated CLS-toxic to the seeds of cress. The germination of seeds treated with CBC № 1, 2, 3, 6, 7, 8, 9, 10, 12, 13, 14, 20 was greater than in the control, equal to 85,7 ± 1,2 percent. Maximum germination of seeds—98,0 ± 1,5 percent, 95,0 ± 0,6 percent, 96,0 ± 1,4 percent, was observed in the processing of their communities of cyanobacteria № 7, 14, 20, respectively (Table 11.1).

Analysis of the data showed that the activity have growth-stimulate CBC № 2, 5, 6, 7, 11, 12, 14, 15, 16, 21. The inhibitory effect was observed in twelve CLS. The community number 24 there was a pronounced inhibitory effect, the average length of the root of six times and the average length of the stem is 1.6 times lower than in controls. Neutral effect was observed in seeds exposed bacterization CBC № 1, 4, 13. Some communities have shown stimulating activity relative to seedling root and stem growth suppressed, and vice versa (CBC № 6, 10).

The highest activity showed growth-stimulate community of cyanobacteria № 5, 11, 14, and 21. The average length of seedlings treated with CBC data, exceeded the control variant in the range of 5.8 to 18 mm. According to the literature it is known that growth-stimulate effect of cyanobacteria associated with the presence of auxin and gibberillin alike substances.

Thus, as a result of the experiment, selected CBC № 2, 5, 6, 7, 11, 12, 14, 15, 20, 21, which can be used for further experiments, including field, with plants growing in the Astrakhan region, as well as for developments in agricultural biotechnology.

KEYWORDS

- **Astrakhan region**
- **Cyanobacteria**
- **Cyanobacterial community**
- **Phyto- and growth-stimulate activity**
- **Soil algae**

REFERENCES

1. Bataeva, U. V.; Dzerzhinskay, I. S.; and Mwali Kamukvamba; South of Russia: The Environment, Development. **2010**, *4*, 76–78.
2. Gollerbah, M. M. et al.; Blue-green algae. Determinant of freshwater algae of the USSR, Moscow, **1953** (in Russian).
3. Gollerbah, M. M.; and Shtina, E. A.; Soil Algae. Nauka, Leningrad, **1969**. 228 p. (in Russian).
4. Goryunova, S. V.; Rzhanova, G. N.; and Orleanskiy, V. K.; Blue-green algae (biochemistry, physiology, role in the practice). Nauka, Moscow, **1969**. (in Russian).
5. Zenova, G. M.; and Shtina, E. A.; Soil Algae: Tutorial. Moscow state university, Moscow, **1990;** 80 p. (in Russian).
6. Pankratova, E. M.; Trefilova, L. V.; Zyablyh, R. Y.; and Ustyuzhanin, I. A.; *Microbiology*. **2008**, *77*(2), 266–272.
7. Pankratova, E. M.; Success of Microbiology, **1987**; 241–242.
8. NETRUSOV A. I.; (Ed); Workshop on Microbiology. Academia, Moscow, **2005;** 352 p. (in Russian).
9. Shtina, E. A.; Zenova, G. M.; Manucharova, N. A.; *Soil. Sci.* **1998**, *12*.
10. Shtina, E. A.; and Hollerbach, M. M.; Nauka, Moscow, **1976**;143 p.

CHAPTER 12

A RESEARCH NOTE ON HEAVY METALS CHANGE EFFECTS OF PIRACETAM ON LEARNING AND MEMORY

O. V. KARPUKHINA[1*], K. Z. GUMARGALIEVA[1], S. B. BOKIEVA[1], and A. N. INOZEMTSEV[2]

[1]N. N. Semenov Institute of Chemical Physics, RAS, 4 Kosygin Street, Moscow, Russia
[2]M. V. Lomonosov MSU, Biological Faculty, Leninskie Gory, 119991 Moscow, Russia
*E-mail: olgakarp@newmail.ru

CONTENTS

A negative influence of heavy metals on many systems of the body including CNS is well known. Employees in the industry of heavy metals suffer from neurodegenerative disorders, including Alzheimer's and Parkinson's diseases [1]. Children who live in industrial cities exhibit signs of retardation of neuropsychic development, including memory, learning, motor function, and speech, as well as a decrease in IQ and other alterations [2–4]. Neuropsychotropic drugs, particularly nootropics, which enhance learning and memory, are widely used [5].

Studies on the effects of heavy metals and nootropic drugs on the CNS have been performed independently from one another, though combined effects of these substances could be supposed. Specifically, the antioxidant system of the body may be a target for combined action of heavy metals and nootropics. The antioxidant system plays an important role in the response of the CNS to stress, because it prevents excessive activation of free radical oxidation processes, which damage cell membranes and contribute to progression of pathologies that influence many functions. Cells of the CNS are the most vulnerable to free radical processes [6–10]. It has been reported recently that neurotoxic effects of heavy metals are related to their ability to induce lipid peroxidation and change the activity of membrane-bound enzymes [8, 11]. On the other hand, the mechanism of nootropic action is determined by their antioxidant and membrane-protective effects [5, 7].

Melatonin, a hormone of the pineal gland with nootropic properties [12], decreases free radical damage of neurons induced by lead [13]. The above data suggest combined effects of nootropics and heavy metals on the antioxidant system of the brain. The purpose of this study was to investigate the effects of the reference nootropic drug piracetam and salts of heavy metals on the conditioning of active avoidance response based on electric footshock reinforcement.

This response was used as an animal model of learning and memory. Fourteen groups of male white nonlinear rats weighing 160–200 g at the beginning of the experiment were used for the study. Group 1 was used as a control for groups 2–10, and group 11 was used as a control for groups 12–14 (table). Groups 1–10 consisted of 11 rats each and groups 11–14 consisted of 12 rats each. The experimental chamber was divided into two equal compartments by a wall with a hole. Avoidance conditioning was performed using the following protocol. Each training trial began with 10-s presentation of an acoustic conditioned stimulus followed by electric footshock (0.5–0.7 mA) through the electrified floor of the chamber. The stimuli were presented in the compartment where the rat was located. If the rat did not enter the safe compartment of the chamber during 10 s, both stimuli were turned off. The stimuli were presented again after a 30-s interval. If the rat entered the adjacent compartment during the sound presentation, it avoided the footshock, and the sound was turned off. If the rat entered the adjacent compartment during the footshock, both stimuli were turned off. Each training session consisted of 25 presentations of the stimuli. During five experimental days, the rats were intraperitoneally injected with one of the solutions of heavy metals and a solution of piracetam at a dose of 300 mg/kg 4 h and 0.5 h prior to avoidance conditioning, respectively. Water solutions of heavy metal salts were used at the following concentrations: 10–7 M lead diacetate, 10–4 M cobalt sulfate, 10–7 M cadmium chloride, and 10–5 M ammonium molybdate. Control animals were injected with an equal volume of the solvent 0.5 h prior to training. The results are presented as the mean numbers of avoidance responses and standard errors of the means. The time courses of avoidance learning in groups were analyzed using the one-way nonparametric Kruskal–Wallis test. The differences between groups were estimated using Wilcoxon's test.

The data presented in the Table 12.1 demonstrate that the mean numbers of avoidances in the animals treated with piracetam were equal to, or even significantly higher than, those in the control animals. These changes were observed in animals of the second group on the second and third experimental days. This is in accordance with the known data on the influence of piracetam on learning and memory under normal conditions [5, 14, 15].

TABLE 12.1 Effects of piracetam and salts of heavy metals on avoidance response formation

Group	Substance	n	Experimental days				
			1	2	3	4	5
1		11	11,6±3,6	25,5±4,8	47,6±3,3	67,3±4,3	85,5±1,8
	Solvent						
2	Piracetam	11	20,4±4,5	41,1±4,4*	61,5±1,8**	72,9±2,4	89,1±1,8
3	Lead diacetate	11	13,5±5,6	21,8±6,1	27,3±6,6**	36,7±5,7**	32,4±4,5**
4	Cobalt sulfate	11	6,2±1,7	16,7±3,0	16,2±3,6 ***	13,1±3,7 ***	10,2±2,9**
5	Cadmium chloride	11	8,4±4,5	17,1±5,4	15,6±5,5 ***	14,2±5,9 ***	14,9±4,2 ***
6	molybdate of ammonium	11	25,8±6,9	36,7±6,3	42,2±7,5	60,7±7,3	64,4±9,7*
7	Lead diacetate + Piracetam	11	3,6±1,7*, +	12,7±2,6*, +++	29,1±5,3** +++	39,6±7,3** +++	43,3±7,0 ***, #, +++

TABLE 12.1 *(Continued)*

Group	Substance	n	Experimental days				
			1	2	3	4	5
8	Cobalt sulfate +Piracetam	11	12,4±1,7	20,0±2,2 +	32,0±3,8* ##, +++	32,2±5,4 ***, ##, +++	44,0±4,7 ***, ###, +++
9	Cadmium chloride + Piracetam	11	7,3±2,1 +++	15,0±3,9 +++	17,6±3,1 ***, +++	26,9±2,0** ##, +++	22,6±4,3** +++
10	molybdate of ammonium+ Piracetam	11	22,6±4,5	34,6±5,2	52,7±5,4	58,6±5,8 +	71,1±5,2+
11	Solvent	12	12.0±3.2	36.7±2.9	61.3±2.9	79.0±1.7	92.6±0.9
12	Piracetam	12	21.7±5.1	40.0±4.6	59.0±1.9	73.0±3.4	89.0±1.6
13	Cadmium chloride	12	10.7±4.3	15.3±5.1 **	14.0±5.2 ***	17.3±6.1 ***	15.7±5.1 ***
14	Cadmium chloride + Piracetam	12	3.7±1.6 **, ++++	12.0±3.6 ***, ++++++	12.7±2.7 ***, ++++++	18.3±2.6 ***, ++++++	15.3±3.3 ***, ++++++

Note: Significant differences between the mean values in each experimental group treated with the drug or metal and the respective control group are indicated by *, +, or #. One, two, or three signs mean the first, second, or third level of significance.

Heavy metals inhibited learning in the animals. Salts of cadmium and cobalt had more pronounced effects. Administration of these compounds led to a lower number of footshock avoidances. The increase in the number of avoidance responses was found on the second experimental day only. Later on, the level of electric footshock avoidances did not elevate, possibly due to the cumulative effect of treatment with these metals. The animals avoided the footshock in 10–17 percent of all trials. The Kruskal–Wallis test showed the absence of significant increase in the number of avoidances from session to session, which confirmed a strong inhibition of learning by these metals. The inhibiting effect of ammonium molybdate on learning was the weakest, and a significant decrease in the number of avoidances compared to the control was observed on the last experimental day only. In accordance with the nootropic hypothesis, the effects of piracetam are more expressed when gnostic and mnestic processes occur under adverse conditions.

It has been demonstrated in experiments with animals and in clinical practice that nootropics, including piracetam, are more effective during aging; in amnesia induced by maximal electroshock, scopolamine, or hypoxia; in the cases of functional impairments of avoidance conditioning; and so on [5, 14].

The expected protective properties of piracetam could be estimated by its capacity for preventing or attenuating the inhibition of learning induced by heavy metals. The ability of piracetam to prevent inhibition of avoidance was observed only in rats treated with the molybdenum salt, which had the weakest effect on learning. The number of avoidances in group 10, where the rats were injected with this metal and piracetam, did not significantly differ from the control level on the fifth day only. During the last 2 days of training, the rats treated with ammonium molybdate prior to piracetam exhibited fewer avoidances than the animals injected with the nootropic alone. In experiments with other groups of animals treated with combinations of substances, piracetam only attenuated the inhibiting influence of heavy metals. Piracetam decreased the effect of cobalt sulfate during the last 3 days, and the number of avoidance responses in the rats treated with this metal and piracetam was greater than in the animals injected with the metal salt only. However, avoidances were fewer as compared to the control group. These data show inhibition of learning under these conditions. The number of avoidance responses in the rats treated with cobalt sulfate and piracetam was smaller than in the animals injected with the nootropic only. Lead diacetate, which was injected to rats of group 3, inhibited learning during three days. Piracetam decreased this effect on the last day only, when the level of avoidance responses in the rats treated with this metal and the nootropic was greater as compared to the animals injected with the metal alone. However, this level was lower than in the control animals. On this day, as well as on others, lead diacetate attenuated the protective action of piracetam, and the level of avoidances was lower in the animals injected with the combination of substances, in comparison with the rats treated with the nootropic alone. The above-mentioned improvement of learning after piracetam treatment, which was observed on the second and third experimental days, was absent after simultaneous administration of lead diacetate and piracetam. Moreover, administration of both substances additionally decreased the number of avoidances during the first two days of training, and the level of these responses remained lower than that in the control group during all 5 days of training. This effect was not found in the group treated with the metal alone. Piracetam decreased the inhibition of avoidance learning in rats of group 9, which were injected with cadmium chloride, on the fourth day only. Though the number of avoidances was higher after treatment with both substances as compared to treatment with the metal alone, it was 2.5 times smaller than in the control group. Throughout the period of training, the effect of piracetam was attenuated by preliminary administration of cadmium chloride, so that the numbers of avoidances in groups 9 and 14 were lower compared to groups 2 and 12, which were injected with the nootropic alone. Moreover, simultaneous administration of cadmium salt and piracetam to the rats of group 14 decreased the number of avoidances compared to the control on the first training day. This effect was not found in group 13, which was treated with the metal alone.

These data demonstrate that piracetam protected from the inhibitory effect of heavy metals on learning and memory. However, the protective effect of piracetam was attenuated by heavy metals in many cases. This inability of piracetam to prevent the inhibitory effect of heavy metals contradicts the results of many studies that demonstrated the capacity of nootropics for recovering impairments of learning and memory. Moreover, the inhibition of avoidance was even more expressed after administration of a nootropic drug together with salts of lead and cadmium as compared to the treatment with these metals without piracetam. Our data show that the use of nootropics is dangerous in the regions with high levels of heavy metals, because it may aggravate the neurotoxic effects of the metals on the central nervous system of humans.

KEYWORDS

- **Avoidance**
- **Heavy metals**
- **Learning and memory**
- **Piracetam**

REFERENCES

1. Gorell, J. M.; Johnson, C. C.; and Rybicki, B. A.; et al. *Neurotoxicology.* **1999**, 20, no. 2/3, 239–247.
2. Doklad o svintsovom zagryaznenii okruzhayushchei sredy Rossiiskoi Federatsii i ego vliyanii na zdorov'e naseleniya (Belaya kniga) (Report on the Lead Environmental Pollution in the Russian Federation and Its Effect on Population Health: The White Data Book), Snatkin, V. V., Ed., Moscow: REFIA, **1997**.
3. Revich, B. A.; and Sidorenko, V. N.; Metodika otsenki ekonomicheskogo ushcherba zdorov'yu naseleniya ot zagryazneniya atmosfernogo vozdukha. Posobie po regional'noi ekologicheskoi politike (Methodology of Assessment of the Economic Damage to Population Health Due to Air Pollution: A Manual on Regional Environmental Policy), Moscow: Akropol', **2006**.
4. Gorobets, P. Yu.; Il'chenko, I. N.; Lyapunov, S. M.; and Shugaeva, E. N.; Prof. Zabol. Ukrepl. Zdor. **2005**, *1*, 14–20.
5. Voronina, T. A.; and Seredenin, S. B.; *Eksp. Klin. Farmakol.* **1998**, *61*(4), 3–9.
6. Aleksandrovskii, Yu. A.; Poyurovskii, M. V.; and Neznamov, G. G.; Nevrozy i perekisnoe okislenie lipidov (Neuroses and Lipid Peroxidation), Moscow: Nauka, **1990**.

7. Dyumaev, K. M.; Voronina, T. A.; and Smirnov, L. D.; Antioksidanty v profilaktike i terapii patologii TsNS (Antioxidants in the Prevention and Treatment of CNS Pathology), Moscow: Nauka, 1995.

8. Zozulya, Yu. A.; Baraboi, V. A.; and Sutkovoi, D. A.; Svobodnoradikal'noe okislenie i antioksidantnaya zashchita pri patologii mozga (Free Radical Oxidation and Antioxidant Protection in Cerebral Pathology), Moscow: Znanie, **2000**.

9. Pshennikova, M. G.; in Aktual'nye problemy patofiziologii (izbrannye lektsii) (Current Problems in Pathophysiology: Selected Lectures), Moscow: Meditsina, **2001**, 220–353.

10. Kutlubaev, M. A.; and Farkhutdinov, R. R.; Akhmadeeva, L. R.; and Mufazalov, A. F.; *Byull. Eksp. Biol. Med.* **2005**, *140*, 10, 414–417.

11. Flora, G. J. and Seth, P. K.; Cytobios, **2000**, *103*, 103–109.

12. Arushanyan, E. B.; In Sovremennye aspekty khronofiziologii i khronofarmakologii (Current Trends in Chronophysiology and Chronopharmacology), Stavropol': Stavr. Gos. Med. Akad., 2004, 9–36.

13. El-Sokkary, G. H.; Kamel, E. S.; and Reiter, R. J.; *Cell. Mol. Biol. Lett.* **2003**, 8(2), 461–470.

14. Inozemtsev, A. N.; Consistent Patterns of Disturbance and Correction of Various Forms of Animal Behavior Using Neuropsychotropic Drugs, Extended Abstract of Doctoral (Biol.) Dissertation, Moscow: Mosk. Gos. Univ., **1997**.

15. Pohle, W.; Becker, A.; Grecksch, G., et al., *Seizure.* **1997**, *6*(6), 467–474.

CHAPTER 13

INVESTIGATION ON THE ACTION OF TWO TYPES OF BIOLOGICAL ACTIVE SUBSTANCES TO THE SOLUBLE PROTEINS THAT ENRICHED THE ANIMAL'S BLOOD SERUM

O. M. ALEKSEEVA[1] and YU. A. KIM[2]

[1]Emanuel Institute of Biochemical Physics, Russian Academy of Sciences, Moscow, Kosygin, 4, Moscow, 119334, Russia; Fax: (499) 137-41-01; E-mail: olgavek@yandex.ru

[2]Institute of Cell Biophysics, Russian Academy of Sciences, Pushchino, Moscow region, Russia

CONTENTS

13.1   INTRODUCTION

The main goal of this work was the investigation of the actions of two types of synthetics biological active substances to the soluble proteins that enriched the animal's blood serum. The first task was the test of the influence of plant growth regulator, applied in agriculture,—Melafen, to the structural properties of soluble protein—bovine serum albumin (BSA). The second task was the investigation of the influence of hybrid antioxidants IHFANs that was expected to use as neuroprotectors, to the BSA structure properties

The first targets are the blood cells and the components of blood plasma also, when biological active substances appeared into the blood-vascular system. This is why we carried out the serum albumin as the test for investigations of biological active substances actions. Serum albumins are water soluble globular proteins. BSA is a simple model of primary serum targets. The serum albumin has small size and, as the component that enriched blood plasma up to 50 percent, plays the essential role in the maintaining of osmotic balance sheets. As albumin enriches the blood sera, it facilitates the correct distribution of tissue liquid at many cases. The total area of all surfaces of albumins is biggest thanks to large amount of molecules and little molecular size of albumin. Besides this, the molecule may adsorb as hydrophilic, and the hydrophobe materials. These is why, the albumins are high effective carriers of most different molecules in blood plasma. And BSA takes essential part in transport of fatty acid, vitamins, hormones and other materials that are needed to animal's organism good functioning.

So, the main purpose of our work was to determine how the aqueous solutions or emulsions of two types of synthetic biological active substances: Melafen and IHFANs, in a wide range of concentrations influence to the structure of soluble proteins with animals originated. Due to the present work the Melafen and IHFANs, action on soluble proteins has been examined. By this obtained data we may suppose or even predict how the biological active substances will be influenced on the protein's structure as well as protein-lipid interactions. Both these parameters are very important for a number of biological functions of albumins at animal body. So, if Melafen or IHFANs have some successful impacts on albumin molecular structure, which may be following by certain, changes of albumin properties, too. At this case, the transport function, or ability to support the osmotic balance will be changed, too.

Thus the selection of BSA as of experimental object was determined by the number of causes. Albumins have the famous structural and functional properties. The serum albumin is 50 percent from mass of all containing in blood sera of proteins. This is one of the first targets for biological active substances in blood serum composition. BSA is the perfect carrier for a numerous materials:

endogenous ones, like some free fatty acids, hormones, metal ions, bilirubin and etc., and exogenous ones (e.g., materials that we want to test). Its structure is labile, and varies very easily. The molecular interaction of serum albumins with transported materials is determined of albumin's structural mobility, conditioned by the loop's stowage of one polypeptide chain, composed of 582 amino acid residues (Figure 13.1). Polypeptide chain forms 9 loops that are fixed by 17 disulfide bonds. It is assumed that the polypeptide chain is laid in three more or less independent cooperative domains. One free SH-group exists in albumin molecule, which can take part in education of disulfides. Disulfides are at the core of trigger assembly of denaturation of this protein.

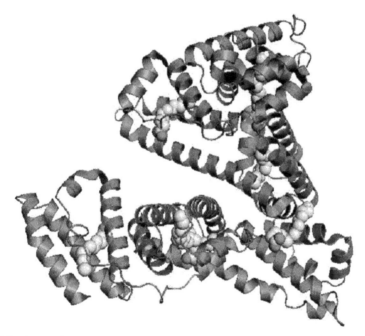

FIGURE 13.1   Scheme of serum albumin molecular structure.

Some changes of serum albumins conformation were registered on change of extent of quenching its intrinsic fluorescence. The numerous works is performed by this time, witch using of this approach for the test of actions of any biological active substances on albumins [1]. The albumin's binding with the exogenous synthetics materials we tested by using the registration of the intensity of intrinsic fluorescence of the BSA. BSA contains 2 tryptophane residues in hydrophobic regions of its molecule. There is the fluorescent emission of 2 tryptophane residues in hydrophobic regions of molecule BSA after excitation of tryptophane. First residue is located with close to a surface, second residue located at the deep inside of the protein globule. When the BSA molecule loosens or unfolds the availability of tryptophane residues for quencher, oxygen, which was dissolute in water, increases greatly. The quenching of tryptophane fluorescence was observed at this case. These changes of BSA tryptophane fluorescence intensity we registered with or without Melafen or IHFANs, when the wide concentration's region. And on the base of these data we may conclude, what Melafen- or IHFANs-aqua solutions or emulsion, under the certain concentration's region, can influence to the BSA structure. Than we may suppose how our tested material and under what region of concentration may have any influence to the functional properties of BSA.

The first task was the test of the influence Melafen to the structural properties of soluble proteins—bovine serum albumin (BSA). Melafen is a plant growth regulator—heterocyclic organophosphor compound, synthesized at the A. E. Arbuzov Institute of Organic and Physical Chemistry of RAS. Melafen is the melamine salt bis (oximethyl) phosphinic acid. It was acquired by one stage with high stepping out of industrially available products [2]. Melafen is a hydrophilic poly functional substance (Figure 13.2).

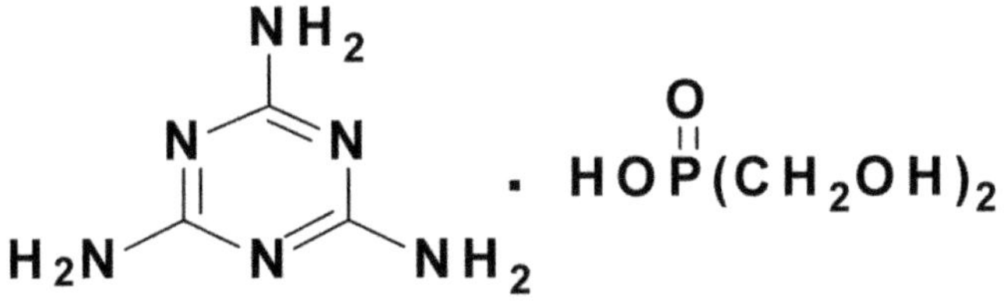

FIGURE 13.2   The structural formula of Melafen molecule.

Melafen raises the plants stress-resistance in the conditions of overcooling and drought, increasing the effectiveness of energy metabolism. In this case Melafen causes the change of the fatty acid composition and the microviscosity of microsome and mitochondrial membranes in vegetable cells [4, 5]. Melafen is the strong regulator of plants stress tolerance under the bed environment. Aqueous solutions of Melafen at concentration $10^{-11} - 10^{-9}$ M increased the plant growth, but the raising of concentration of Melafen up to 10^{-8}, 10^{-7} M leads to plant's seeds dye. Therefore, our studies were carried out in a wide range of concentrations ($10^{-21} - 10^{-3}$ M).

Taking into account the close interdependence of plant's and animal's bodies in nature, it was necessary to investigate the action of plant growth regulator at any objects of animal origin. The primary targets for biological active substances in animal's cells are membrane and their components. Performed analysis of actions of aqueous solutions of Melafen to the structural and functional characterizations of lipids and protein that built into the cellular membranes [6–8], was complemented by the testing of Melafen influences on free soluble proteins unbounded with membranes. As such model the protein of bovine serum albumin (BSA) was used.

The second task was the investigation of the influence of hybrid antioxidants IHFANs to the BSA structure properties. IHFANs were expected to use as neuroprotector. For the experiments IHFANs were used as the aqua-ethanol suspensions at the wide concentration range (10^{-21} M – 10^{-3} M). IHFANs are the derivatives of antioxidant phenozan (β - (4-hydroxy-3, 5-di-tert-butylphenyl) propionic acid). Phenozan was created for stabilization of polymer at Institute of Chemical Physics of RAS Moscow [9]. It is known that the antioxidants often shall be used for the therapy of any pathological states. The derivative of phenozan—its potassium salt, was tested as biological active materials. It was turned out that potassium salt of phenozan exhibited the property of strong antioxidant and structural effectors on enzymes and on biomembranes [10]. However, the phenozan didn't have the certain targets for its actions at membrane. As the amphiphilic agent, phenozan acts primarily at all regions of surface layers of biomembrane, as in exterior and in internal sheets of bilayer. It was appeared, that phenozan penetrated through biomembrane defects to internal surface of bilayer plasma membrane, that was discovered with using of spin-labeled EPR probes on erythrocyte membranes [11] For ingress into deeper layers of membrane without defects, it was necessary synthesized the more hydrophobic antioxidant. For that purpose the choline esters was added to phenozan that was quaternized by long chain alkyl halogenides with number of carbon atoms from 8 up to 16 - (4-hydroxy-3, 5-di- tert- butyl phenyl) propionyl butyl] ammonia halogenides that were named IHFANs [12].The series of hybrid multitarget antioxidants—IHFANs for orientation of antioxidant action, had been synthesized in IBHF RAS. Constructions were biological active. It based on phenozan with conservation of screened phenol. And one choline residue and one alkyl residues of different lengths (C8-C16) were inserted in that the complex molecules. So that molecules received antioxidant activity, and bought the new activities: the anticholinesterase activity, and IHFANs received the ability to penetrate bilayer by introducing of alkyl residues of different length (C8-C16) into the hydrophobic regions [13]. Scheme of IHFANs is presented at Figure 13.3.

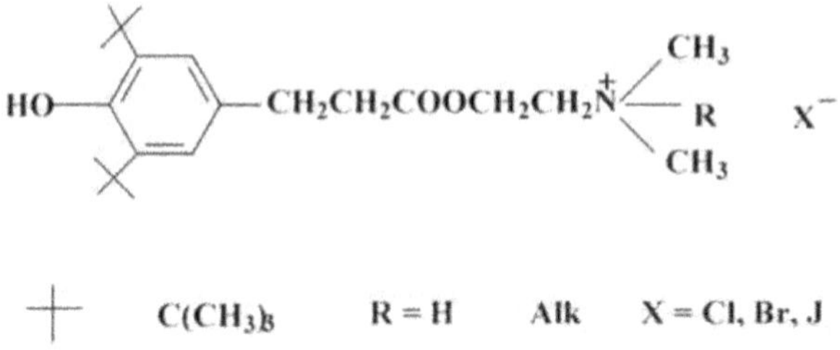

FIGURE 13.3   The structural formulas of IHFANs molecules. (C8) R = C_8H_{17}; (C10) R = $C_{10}H_{21}$; (C12) R = $C_{12}H_{25}$; (C16) R = $C_{16}H_{33}$; X = Br-.

As was shown at Figure 14.3, the hybrid antioxidants—IHFANs, have a charged onium group and a lipophilic long-chain alkyl tail. These structures of complex molecules allowed them to interact effectively with a charged lipid bilayer, insert to hydrophobic regions of cell membranes and maintain the antioxidant status. These molecules are bounded on membrane surface by the positive charge on quaternary nitrogen (the anchor), and the alkyl residue (the float) is introduced in bilayer being disposed among fatty acid's residues of phospholipids. And alkyl halogenides with variable length, that were added to phenozan: R-C8H17; C10H21; C12H23; C16H33, were implemented to the membrane bilayer on the different deeps. Thus, ICHFANs localized in anion heads regions and in fatty acids tails. This phenomenon fortifies the membrane structure so much. And the membrane became resistant to any bad environment actions.

Performed analysis of actions of aqueous suspension of IHFAN-C-10 to the structural and functional characterizations of protein, that built into the cellular membranes, the erythrocytes, and its ghost [14, 15], were complemented by the influence testing of ICHFANs on free soluble proteins unbounded with membranes. As such model was used the protein of bovine serum albumin (BSA). These experiments of IHFANs influences to BSA structure, were similar as experiments of Melafen testing were performed with aid of quenching if intrinsic fluorescence method.

13.2   MATERIALS AND METHODS

The materials: BSA (Sigma). Melafen (melamine salt bis (oximethyl) phosphinic acid) was synthesized at the A. E. Arbuzov Institute of Organic and Physical Chemistry of RAS Kazan. IHFANs (4-hydroxy-3, 5-di- tert- butyl phenyl) propionyl butyl ammonia halogenides were synthesized at Institute of Biochemical Physics of RAS Moscow.

BSA was used as the aqua solutions. Melafen was used as the aqua solutions at the wide concentration range (10^{-21} M – 10^{-3} M). IHFANs were used as the aqua-ethanol suspensions at the wide concentration range (10^{-21} M–10^{-3} M).

The standard methods with the standard conditions had been used for measurements of fluorescence quenching: 1mkM of BSA protein aqua solution, 20°C. The quartz cell (1 sm.) was used for BSA (with or without Melafen or IHFANs) fluorescent intensity registrations by fluorescent spectrophotometer "Perkin-Elmer MPF-44B". The spectrophotometer "Specord M 40" was used for measurements of optical density of BSA protein aqua solutions when low concentrations.

13.3   RESULTS AND DISCUSSION

The first task of this work was the supporting of the Melafen influence to the albumin structure. Melafen at this case was the one of the factor of the certain variable bed environments. It is clear that the probability of albumin molecules to release and absorb the fatty acids and another absorbed substances was under the strong influence of environment. Melafen increase the crop producing power of vegetables and seeds. The plant cells drastically increased of metabolism upon the treatment of small doses of Melafen that followed to greet elevating of plant's resistance to difficult environment. For resolve of the Melafen-BSA interrelation we provided the spectral analysis. Data of the dependence of optical density from varied concentrations of Melafen are presented at Figure 14.4.

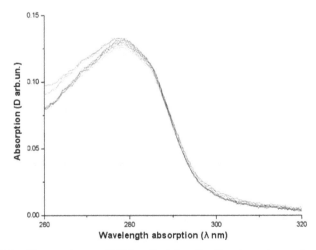

FIGURE 13.4   The spectrums of BSA optical density. The dependence of optical density from varied concentrations of Melafen.

As was shown at Figure 13.4 the effect of aqueous solutions with varied concentrations of Melafen on spectrum patterns of BSA the changes in absorption spectrums of BSA were negligible. The shape of BSA absorption spectrums, the locations of maximums of absorption didn't change drastically, when concentrations of Melafen were varied. Maximum of absorption spectrum BSA didn't not shift, however occurred the some change of the absorption degree and of spectrum shapes. Data of the dependence of correlation of BSA optical density with Melafen to without Melafen (D_{mel}/D) from Melafen concentrations were presented at Figure 13.5. The correlation (D_{mel}/D) was under the polymodal small changing, when Melafen concentrations varied.

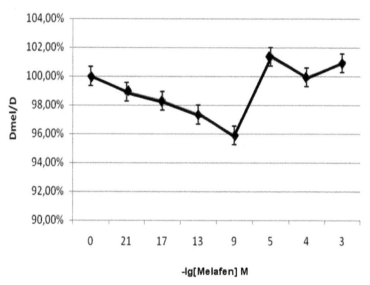

FIGURE 13.5   The dependence of correlation of BSA optical density with Melafen to without Melafen from Melafen concentrations.

Data of Melafen influence to the BSA absorption spectrums, which are presented at Figures 13.4 and 13.5, provides the evidence about absence of the covalent linkage between molecule Melafen and protein BSA. However, in registration of fluorescence spectrum, presented at Figure 13.6, revealed facts, indicative of great influence of aqueous solution of Melafen over a wide range of concentrations on conformation of BSA molecule. All solutions under the different concentrations of Melafen were not displaced the wavelength of fluorescence maximum. However the fluorescence intensity has undergone a change. The great quenching of tryptophan fluorescence, when 10^{-4} M of Melafen was found, and the burst of fluorescence intensity, when 10^{-17}–10^{-10} M was found too (Figures 13.6, 13.7). The changes of fluorescence tryptophan intensity residues BSA are shown at Figure14.6.

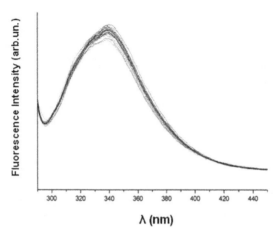

FIGURE 13.6   The influence of Melafen over a wide range of concentrations to the intensity of emission spectra of BSA tryptophan residues.

The data that fluorescence of BSA tryptophane residues was quenched by Melafen under the wide region of concentration were presented at Figure 13.7.

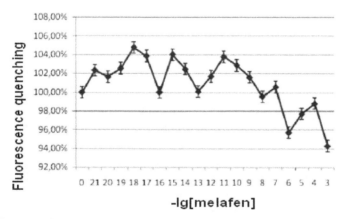

FIGURE 13.7   The Melafen influence to the fluorescence intensity of BSA. The control samples in absentia of Melafen have been adopted the fluorescence intensity as 100 percent. More 100 percent—is some burst of fluorescence intensity, less—is the quenching of fluorescence intensity.

As it can be seen from comparison of data, presented at Figure 14.6, the shapes of emission spectra were similar, only the maximal value of fluorescence intensity was change in dependence from the Melafen concentrations. Then, we build a curve of the dependence of value at the maximum of tryptophan emission from Melafen concentration Figure 13.7. We obtained the noticeable tendency of the fluorescence quenching by the Melafen, when large concentrations. And the some increasing of fluorescence intensity was occurred when low and ultrasmall concentrations of Melafen presented at the experimental medium. The dose-dependence was polymodal, which is representative for biological active substances, effectual in small and ultrasmall doses [16]. Evidently, conformational rearrangements occurred in BSA molecules. These rearrangements were small, and had the different direction.

The second task of our work was the investigation of the influence of hybrid antioxidants IHFANs that was expected to use as neuroprotector, to the BSA structure properties. IHFANs were used as the aqua-ethanol suspensions at the wide concentration range (10^{-21} M – 10^{-3} M). The spectral analysis of BSA-IHFANs interrelationships was occurred similar as for testing of Melafen–BSA actions.

The soluble protein the serum albumin (BSA) in the presence of large concentrations of testable materials was turning, and becomes greatly accessible to water introduction. In the presence of low concentrations, on the contrary, the protein structure appears to much be getting stronger. The alkyl-halogenides tails of IHFANs are adsorbed on protein and defend it from molecular untwisting and water ingestion. The degree of protection was depended on length of alkyl tail directly in proportion. The maximum of protection has been found for IHFAN-C16 relationships with BSA (Figure 13.8).

The proteins, which tryptophanes are contained, absorb the light near 280 nm, and the fluorescence spectrums are shifted to the short wavelength in contrast with tryptophane spectrums in water. Respectively importance their maximums may change from 343 nm (the serum albumin). Primary cause of this is the processes presence of orientation interaction, in which the spectrum regulation is determined by the polarity and microenvironment rigidity of chromophore. The dislocation of a high of the short wavelength is common to no polar environment, in long-wave—for polar. The emitting maximums of proteins reflect average the availability their tryptophane residues in aqueous phase.

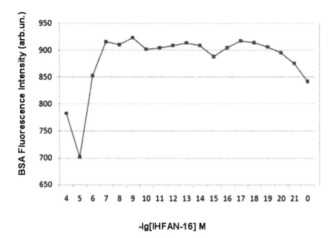

FIGURE 13.8   The influence of IHFAN-C16 to the tryptophan fluorescence intensity of BSA Dependence of fluorescence intensity BSA from concentration IHFAN-C16 (10^{-21}–10^{-3} M).

As was shown at Figure 13.9 the emitting maximums of BSA were shifted under the varied IHFAN-C16 concentrations. This shift was bigger when IH-FAN concentrations were 10^{-6} M–10^{-3} M. Are likely IHFANs "stick all over" the molecule of BCA, and as result the environment of tryptophane residues becomes more hydrophobe, than at IHFANs absences or when low their concentrations existance.

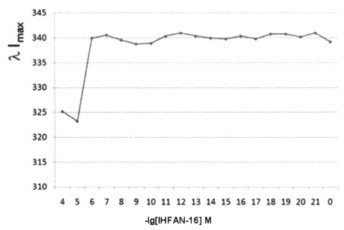

FIGURE 13.9   The influence of IHFAN-C16 (10^{-21}–10^{-3} M) to the tryptophane fluorescence intensity of BSA. The dependence of changes of wavelength of BSA fluorescence maximum from concentration IHFAN-C16.

When we test the IHFANs actions at the soluble protein BSA (Figures 13.6 and 13.7), we found the decreasing of tryptophane fluorescence when 10^{-4} M and the raising when 10^{-17}–10^{-7} M. While for IHFAN-C16, which have the greatest alkyl-halogenides tail, the inflammability had been greatest. Are likely, IHFANs communicate with albumin in low concentrations, adsorbed at BSA surface. So IHFANs when low and ultralow concentrations defend tryptophane residues from contacts with the water. And when large concentrations, IHFANs change the BSA structure, are likely, and increasing the availability of tryptophane for water. Some oxygen, which was solute at water, quenched the intrinsic tryptophane fluorescence of BSA molecules. But IHFANs, when the large concentrations (10^{-5} M – 10^{-3} M), not only decreased the intrinsic BSA fluorescence, but IHFANs shifted the fluorescence maximum to more short-wavelength. This indicates the initiation of more polar environment for BSA tryptophane residues.

13.4   CONCLUSION

Evidently, Melafen molecules affected to the albumin so that under the small and ultrasmall concentrations there was the preserving of the protein tryptophane residues from quenching from oxygen, dissolved in water. And under the large Melafen concentrations the change of protein conformation became so essential. At this case the tryptophane residues that lying at deep molecule locus became more available to water (and oxygen, respectively), on that indicated the fluorescence quenching. Occurs "the loosening" of BSA molecule structure. We may conclude that the soluble proteins that unhardened of the membrane lipids were under the essential Melafen actions. Taking into account that Melafen is the hydrophilic substance, and it can change the water environment. At this case we may suppose that Melafen influence to BSA by two ways: mediated through the water, or directly to the influence to hydrophilic sites of BSA molecules. Mechanism was unknown. These influences were mainly changed in dependence on Melafen concentration present in surrounding solution. There were not clear evidences of BSA-Melafen linkage existence. However mediated action through the change of water medium appears to occur surrounding the protein's molecules.

Also the water solutions of Melafen may be the regulator of transporting function of albumins, as its will be introduced to the animal's body. And it may be take part in extracting fatty acids from any molecules, or bounding of free fatty acids by albumins. As its known the water solutions of Melafen change the fatty acids content of membranes [5].

The IHFANs actions on the BSA structure were different from Melafen actions. Evidently, because the Melafen—is the simple neutral hydrophilic substances, but the IHFANs are the complex substances with hydrophobic part—long chain alkyl halogenides, and hydrophilic part—positive charge on quaternary nitrogen. By this molecules of IHFANs are bounded on BSA surface by the positive charge and introduced to the deep locus of BSA by chains alkyl halogenides. As was shown at our work, the IHFANs preserve BSA molecule from "loosening" of molecule BSA structure. And the region of this nondestructive IHFANs concentrations was extended more ($10^{-21} - 10^{-7}$ M), than for Melafen nondestructive concentration's region ($10^{-21} - 10^{-9}$ M). The degree of protection was depended on length of alkyl tail directly in proportion. The maximum of preserving has been found for IHFAN-C16. It may be supposed that alkyl halogenides tails what having occupied all hydrophobic regions in albumin molecule, in practice came to form the containment shell from water encroachment to tryptophane residues. Respectively and the oxygen, which was dissolute in water, was failed to penetrate to tryptophane residues. Thus the mechanisms of BSA-Melafen or BSA-IHFANs relationships were different. There were not the covalent binding, but the grade of absorption, points for absorptions were different. The "loosening" of molecule BSA structure occurred, when the large concentrations (10^{-5} M – 10^{-3} M) of Melafen were presented at medium for intrinsic BSA fluorescence registration. And IHFANs, when the large concentrations (10^{-5} M – 10^{-3} M), not only decreased the intrinsic BSA fluorescence. IHFANs shifted the fluorescence maximum to more short-wavelength. This indicates the initiation by IHFANs of more polar environment for BSA tryptophane residues.

Obtained data in this work suggests to the fact that one of the first targets in blood, in particular, in blood plasma, was exposed to as Melafen, and IHFANs, used over a wide range of concentrations. The albumin structure was varied under the presence of Melafen or IHFANs. Respectively, the presupposition was arising that if some properties of albumin change, as carrier of biologically active substance and as osmotically active substance. Albumins maintain the colloid-osmotic pressure at blood plasma and at other fluids (e.g., in cerebrospinal fluid). It can be assumed that the transport effectiveness will be decreased and can be unbalanced of osmotic pressure in compartments with biological fluids. This is why the application these materials demands the great cares and the observances of concentration limitations, because the soluble protein of animal origin—bovine serum albumin, changes his structural properties in their attendance so much.

KEYWORDS

- **Bovine serum albumin**
- **Fluorescence**
- **Hybrid antioxidant**
- **Melafen**

REFERENCES

1. Diaz, X.; Abuin, E.; and Lissi, E.; Quenching of BSA intrinsic fluorescence by alkylpyridinium cations its relationship to surfactant-protein association. *J. Photochem. Photobiol.* **2003**, *55*, 157–162.
2. Fattachov, S. G.; Reznik, V. S.; and Konovalov, A. I.; Melamine Salt of Bis (hydroxymethyl) phosphinic Acid. (Melaphene) As a New Generation Regulator of Plant growth regulator. In set of articles. Reports of 13th International conference on chemistry of phosphorus compounds. S. Petersburg. **2002**. S. 80.2.
3. Kostin, V. I.; Kostin, O. V.; and Isaichev, V. A.; Research results concerning the application of Melafen when cropping. In Investigation State and Utilizing Prospect of Growth Regulator "Melafen" in Agriculture and Biotechnology, Kazan. **2006**, 27–37.
4. Zhigacheva, I. V.; Fatcullina, L. D.; Burlakova, E. B.; Shugaev, A. I.; Generosova, I. P.; Fattahov, S. G.; and Konovalov, A. I. Influence of phosphoorganic plant growth regulator to the structural characteristics of membranes plant's and animal's origin". *Biol. Membr.* **2008**, *25*(2), 150–156.
5. Zhigacheva, I. V.; Misharina, T. A.; Trenina, M. B.; Krikunova, N. N.; Burlakova, E. B.; Generosova, I. P.; Shugaev, A. I.; and Fattahov, S. G.; Fatty acid's content of mitochondrial membranes of Pea seedlings in conditions of insufficient moistening and treatment by the phosphoorganic plant growth regulator. *Biol. Membr.* **2010**, *27*, 256–261.
6. Alekseeva, O. M.; Fatkullina, L. D.; Burlakova, E. B.; Fattakhov, S. G.; Goloshchapov, A. N.; and Konovalov, A. I.; Melafen influence on structural and the functional state of liposomes membranes and cells of ascetic Ehrlich carcinoma. *Bull Exp. Biol. Med.* **2009**, *147*(6), 684–688.
7. Alekseeva, O. M.; Krivandin, A. V.; Shatalova, O. V.; Rykov, V. A.; Fattakhov, S. G.; Burlakova, E. B.; and Konovalov, A. I.; The Melafen-Lipid-Interrelationship Determination in phospholipid membranes". *Doklady Akademii Nauk.* **2009**, *427*(6), 218–220.
8. Alekseeva, O. M.; The Influence of Melafen–Plant Growth Regulator, to Some Metabolic Pathways of Animal Cells". *Polym. Res. J.* **2013**. USA. 7(1). Chapter 6. 15–23.
9. Ershchov, V. V.; Nikiforov, G. A.; and Volod'kin, A. A.; Space screened phenols. M. Chemistry. **1972**, 352 p.
10. Burlakova, E. B.; Goloshchapov A. N.; and Treschenkova, J. A.; Action of low doses of phenosan on biochemical properties of lactate dehydrogenase and membranes microviscosity by the microsome of mouse brain. *Rad. Biol. Radioecol.* **2003**, *3*, 320–323.
11. Gendel, L. J.; Kim, L. V.; Luneva, O. G.; Fedin, V. A.; and Kruglakova, K. E.; Changes of cursory architectonics of erythrocytes under the impact of synthetic antioxidant phenosan-1. *Izvestiya. RAS. Ser. Biol.* **1996**, *4*, 508–512.
12. Nikiforov, G. A.; Belostockaya, I. S.; Vol'eva, V. B.; Komissarova, N. L.; and Gorbunov, D. B.; Bioantioxidants "Of float" type on the biologacal active substancesis of derivative 2, 6

ditertbutil fenol". **2003**. set of articles. Bioantioxidants, Scientific Medical academician. Tyumen, 50–51.

13. Burlakova, E. B.; Molochkina, E. M.; and Nikiforov, G. A.; Hybrid antioxidants. *Chem. Chem. Technol.* **2008**, *2*(3), 163–171.

14. Alekseeva, O. M.; Kim, Yu.A.; Rikov, V. A.; Goloshchapov, A. N.; and Mill, E. M.; Influence of screened phenols on lipids structure, and also soluble and membrane proteins. In "Phenolic Compounds: Fundamental and Applied Aspects, Publishing House "Nauchnii Mir". **2010**. (ISBN) Chapter 1. Phenolic compound: structure, property, biological activity, 116–126.

15. Albantova, A. A.; Binyukov, V. I.; Alekseeva, O. M.; and Mill, E. M.; The investigation influence of phenozan, ICHPHAN-10 on the erythrocytes *in vivo* by AFM method. Modern Problems in Biochemical Physicsew Horizons" **2012**. Ed. by S. D. Varfolomeev, E. B. Burlakova, A. A. Popov, G. E. Zaikov Nova Science Publishers. New York. Chapter 5. pp 45–48.

16. Burlakova, E. B.; Effect of ultrasmall doses. *Vestnik RAS.* **1994**. *64*(5), 425–431.

CHAPTER 14

MECHANISM OF PROTECTIVE ACTION OF ANTIOZONANTS

S. D. RAZUMOVSKY, V. V. PODMASTERYEV, and G. E. ZAIKOV

N. M. Emanuel Institute of Biochemical Physics, Russian Academy of Sciences, 4 Kosigin str. Moscow 119334, Russia; E-mail: chembio@sky.chph.ras.ru

CONTENTS

14.1 INTRODUCTION

Destruction of nonsaturated polymers (mainly, rubbers and rubbers) under the influence of ozone is accompanied by formation of cracks on a surface of products, loss of mechanical durability and destruction. In literature there is enough of researches and some survey works which are well generalizing the main results of researches of ozonic destruction of rubbers and ways of fight against it [1–4]. However, despite a large number of researches yet it wasn't succeeded to achieve effective protection against ozonic destruction. Works on search of new protective systems proceed [5, 6]. In the most part these works have empirical character.

14.2 EXPERIMENTAL RESULTS AND DISCUSSION

The technique of synthesis of ozone, its registration, and calculations of constants of speeds of reaction are given in [7].

Most of the highly effective antiozonants belong to the group if substituted N,N'-phenylenediamines (PPD) and perform a great number of diverse functions. Apart from protecting against atmospheric ozone, these compounds serve as effective antioxidants and antifatigue agents.

Figure 15.1 summarizes the results of dynamic testing of natural rubber protected by various antiozonants [8]. Plotted on the ordinate is the mean crack area (S) per unit area of the specimen, which was used in this study as a measure of degradation:

$$S = \sum_{i=1}^{i=n} \frac{lh}{n} \qquad (14.1)$$

where n is the number of cracks, l is the length of a crack, and h is its depth.

Most of the other properties of the vulcanizates also improve perceptibly. Antiozonants increase the time to cracking, improve the specimen durability [9, 10] minimize creep, and reduce the stress relaxation rate in the specimen [11].

The complexity of the cracking phenomena is matched by that of antiozonant action. Several proposals have been made regarding the mechanism of this protective action in polymers; however, there is only one work in which all possible mechanisms are compared within a single experiment [12].

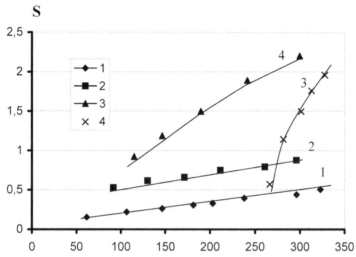

FIGURE 14.1 Effect of test time on the degree of degradation (S) of the surface of naturally derived rubber containing various antiozonants: 1, N,N'-di-1- methylheptil-p-phenylenediamine; 2, N-phenyl-N'isopropyl-p-phenylenediamine; 3, 6-ethoxy-1,2-dihydro-2,2,4-trimetylquinoline; 4, N,N'-dioctil-p-phenylenediamine.

The suggestion that antiozonants serve a catalysts of ozone decomposition [13] cannot be true because ozone react with the antiozonant [14], and the stoichiometry and main products of the reaction are known [15, 16].

Not truth an accept as the proposal [17, 18] that, being bifunctional, antiozonats react with products of interaction between ozone and macromolecules, binding the ends of the disrupted chains, since it is clear that the important objective is not to heal broken chains but to maintain the existing sequences of bonds. Furthermore, monofunctional antiozonants such as tributylthioureas [19] are known as they cannot be broken but are capable of adequately protecting rubbers against atmospheric ozone.

Another hypothesis, which states that antiozonants migrate to the surface where they react with ozone itself or with the products of its reaction with the $C = C$ bonds of the polymer thus forming a film that prevents ozone from penetrating inside toward the polymer, is difficult to verify. A similar effect is to be expected by analogy with waxes. However, there are two indirect arguments against any significant effect of film formation. Firstly, the products of the reaction between ozone and N-isopropyl-N'-phenyl-p-phenylenediamine (4010NA), incorporated into a specimen of cured natural rubber, do not protect it against ozone-induced degradation [16]. Secondly, calculation of the antiozonant diffusion rate indicates that migration toward the surface from the core of the specimen cannot play any significant role during cracking [24].

Nevertheless, judging from many publications [17, 18, 20–23] most researchers believe that the physical properties of antiozonants play an important role in the efficiency of their protective action, although it is not specified which properties. In particular, it seems that only the difference in the physical properties of tributylurea and 4010NA can explain why the times to cracking are comparable only under static conditions, while under dynamic conditions tributylurea is much less effective [27]. Comparison of the behavior of tributyl-thioureas with different substituents suggest that alkyl substituents with an open chain are replaced by cyclic ones, the correlation between the time to cracking and reactivity to ozone remains the same although the curves diverge further (see Figure 14.2) [19].

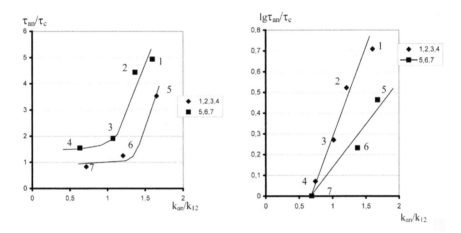

FIGURE 14.2   Ratio of time to cracking of protected specimens (τ_{an}) to that of unprotected control specimen (τ_c) versus ratio between rate constants of the ozone-antiozonant (k_{an}) and ozone-C = C (k_{12}) reactions.

The proposal [17] that antiozonants on the rubber surface compete with the **C = C** bonds for ozone and thereby protect vulcanizates against cracking is adequately corroborated at present. The arguments supporting this assumption have been derived using kinetic methods for studying ozone-macromolecule and ozone-antiozonant reactions in solutions [25, 15, 16]. It was shown that in the presence of antiozonants the rate of macromolecular chain scission slowed down (Figure 14.3)

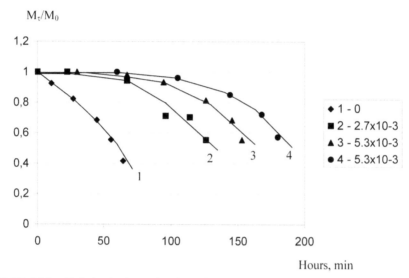

FIGURE 14.3 Variation in the molecular weight of polybutadiene exposed to ozone in a solution at various concentrations of N-phenyl-N'-isopropyl-p-phenilene diamine (mol/liter): 1, 0; 2, 2.7 × 10^{-3}; 3 and 4, 5.3 × 10^{-3}. Molecular weight of polybutadiene: 1–3, 60 000; 4, 4000, CCl$_4$, [O$_3$] = 4 × 10^{-5} mol/liter, 20°C.

1. C$_4$H$_9$NH —C —N(C$_4$H$_9$)$_2$;   2. ⬡ NH —C —N(C$_4$H$_9$)$_2$; (K_{12})
             ‖S   ‖S

3. ⬡ NH—C—N(C$_4$H$_9$)$_2$;
           ‖S

4. O$_2$N ⬡ NH—C —N(C$_4$H$_9$)$_2$;
               ‖S

5. ⬡ NH —C—N(C$_2$H$_5$)$_2$;
           ‖S

6. 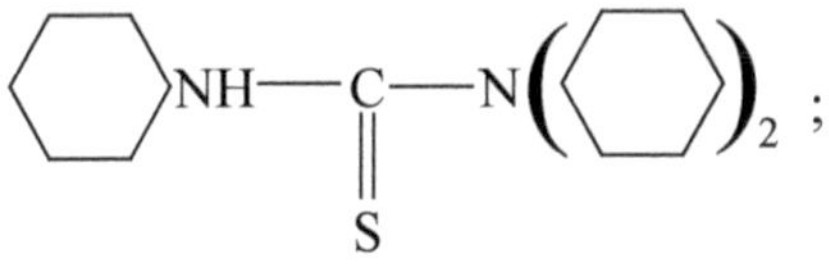 [structure: cyclohexyl–NH–C(=S)–N(phenyl)$_2$] ;

7. 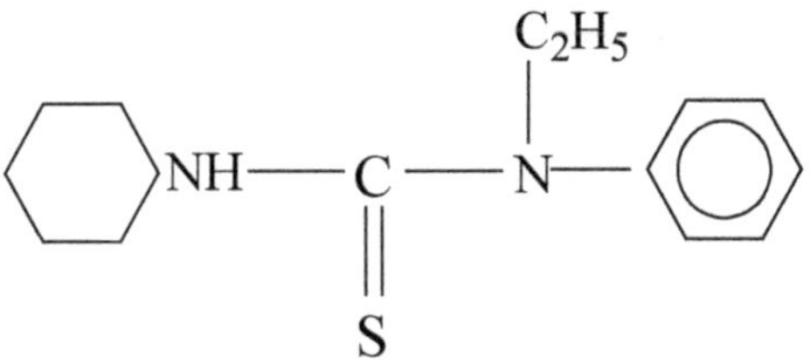 [structure: cyclohexyl–NH–C(=S)–N with C$_2$H$_5$ group and phenyl]

although the absorption of ozone continued. Products of the reaction of ozone with antiozonant accumulated in the solution. Studies into the kinetics of this reaction have shown that it is extremely fast, its rate exceeding those of all presently known reactions involving ozone [26].

Special methods have been developed for measuring the rate constants of the ozone-antiozonant reaction. One method is based on the competition of ozone for a reference compound with a known rate constant and the compound of interest [27], while the other is based on comparison on the ozone dissolution rate and the rate of its reaction [28, 29].

In the first method [27], a thiourea which reacts with ozone in two steps (Scheme 14.1) was selected as the reference compound:

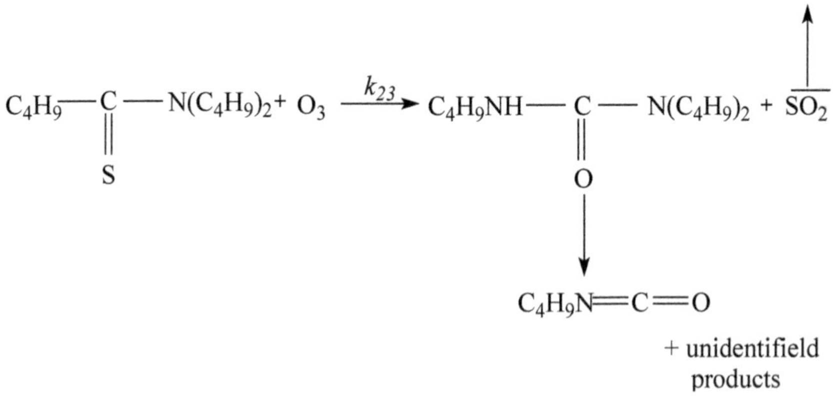

$$C_4H_9-C(=S)-N(C_4H_9)_2 + O_3 \xrightarrow{k_{23}} C_4H_9NH-C(=O)-N(C_4H_9)_2 + SO_2$$

$$\downarrow$$

$$C_4H_9N=C=O$$

+ unidentifield
products

SCHEME 14.1

The first step yields sulfur dioxide. When the reaction is conducted in a bubble reactor, the presence of SO_2 can be detected by absorption in the UV region.

Figure 14.4 represents typical curves for variation in the optical density *(D)* of the mixture at the reactor outlet for the interaction of reference compound, methyl oleate, and tributylthiourea (TBTU) with ozone. Knowing the starting amounts of methyl oleate and TBTU, the gas flow rate and O_3 concentration one can easily calculate the stoichiometric coefficients of the reactions using the area above the curve before ozone appears at the reactor outlet. These were found to be 1 for methyl oleate and 2 for tributylthiourea.

Comparison of the curves representing the optical density of the gaseous mixture at the reactor outlet versus time indicates that the change in optical density at $\lambda = 280–300$ nm (Figure 14.4 (b), 4) differs in shape from that observed at 254 nm, which is representative of the kinetics of the variation in the absorption in the ozone rate (Figure 14.4a). The first step of the reaction between TBTU and ozone yields SO_2 whose absorption maximum can be seen at 285 nm, and this accounts for the increase in optical density during the course of the reaction.

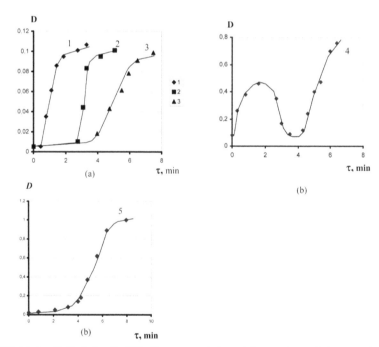

FIGURE 14.4   Optical density (D) of gaseous mixture at the reactor outlet versus time. (a) $[O_3]_0 = 6 \times 10^{-6}$ mol/liter; $1 = 254$ nm; $V_{O2} = 120$ ml/min; 1, CCl_4; 2, 6×10^{-5} mol/liter of methyl oleate in CCl_4; 3, 6×10^{-5} mol/liter of tributylthiourea in CCl_4. (b) $[O_3]_0 = 2 ´ 10^{-3}$ mol/liter; $1 = 290$ nm; $V_{O2} = 120$ ml/min; 4, 6×10^{-3} mol/liter of tributylthiourea in CCl_4; 5, 5×10^{-3} mol/liter of methyl oleate in CCl_4

The resulting SO_2 accumulates in the solution but is partially carried away by the gas flow (Scheme 2)

$$[O_3]_{sol} + TBTU \xrightarrow{k_{21}} [products] + [SO_2]_{sol}$$

$$[SO_3]_{sol} \xrightleftharpoons{\alpha} SO_{2gas}$$

SCHEME 2

where α is the solubility coefficient. In the case the rate of the reaction between TBTU and ozone and the rate of formation of the volatile component depend on the gaseous mixture supply rate and are constant with time $(W_p = \omega[O_3]_0)$, where ω is the specific rate of ozone supply (liter $^{-1}s^{-1}$), whence

$$\omega[O_3]_0\, \alpha\tau = W[SO_2]_{gas}\, \alpha\tau + [SO_2]_{sol} \tag{14.2}$$

or

$$\frac{\alpha[SO_2]_{sol}}{\alpha\tau} = \omega([O_3]_0 - [SO_2]_{gas}) \tag{14.3}$$

Substitution of $\alpha[SO_2]_{gas}$ for $\alpha[SO_2]_{sol}$ and integration gives eqn. (4):

$$\omega[SO_2]_{gas} = \omega[O_3]_0\left[1 - \exp\left(-\frac{\omega\tau}{\alpha}\right)\right] \tag{14.4}$$

where $[SO_2]_{gas}$ is the rate of release of volatile products, $\omega[O_3]_0$ is the rate of the reaction between ozone and TBTU, and $[1—\exp(\omega\tau/\alpha)]$ is the reaction factor of SO_2 in the solution.

When olefin is introduced into the system, the ozone react with both components at the same time (Eqn. (15.5)):

$$\omega[O_3]_{gas} = k_{21}[TBTU][O_3] + k_{12}[olefin][O_3] \tag{14.5}$$

and the participation of TBTU in the total process is given by the expression (6)

$$\frac{k_{21}[\text{TBTU}]}{k_{21}[\text{TBTU}]+k_{21}[\text{olefin}]} \tag{14.6}$$

then

$$\omega[SO_2]_{gas} = \omega[O_3]_0 \frac{k_{21}[\text{TBTU}]}{k_{21}[\text{TBTU}]+k_{12}[\text{olefin}]}\left[1-\exp\left(-\frac{\omega\tau}{\alpha}\right)\right] \tag{14.7}$$

The SO_2 concentration at the reactor outlet will diminish accordingly. The dependence of $[SO_2]_{gas}$ on [olefin] is shown in Figure 14.5. If one uses the ratio $[SO_2]'_{gas}/[SO_2]''$ (D' and D'' are the corresponding optical densities) at different olefin concentrations [olefin]' and [olefin]'' within the same period of time, one can easily determine k_{21} if k_{12}, the rate constant of the reaction between ozone and olefin, is known.

$$k_{21} = \frac{k_{12}[\text{olefin}]'' - \dfrac{D'}{D''}[\text{olefin}]'}{[\text{TBTU}]\left(\dfrac{D'}{D''}-1\right)} \tag{14.8}$$

In the case of tributylthiourea the rate constant of the reaction with ozone (k_{2l}) was found to be $(2\pm0.2) \times 10^6$ liter/mol s.

The rate constants of the reaction of ozone with reference olefin ($k_{21}= 1 \times 10^6$ liter/mol s) and with the product of the first step of the reaction between ozone and TBTU isocyanate ($k_{22}= (3.5 \pm 0.5) \times 10^3$ liter/mol s) were calculated from the slopes of curves 2 and 3 in Figure 14.4(a) using the formula

$$k_{22} = \frac{\omega[O_3]_0 - [O_3]_{gas}}{\alpha[X]\tau[O_3]_{gas}} = \frac{(D_0 - D_\tau)}{\alpha[X]_\tau D_\tau} \tag{14.9}$$

where $[X]_\tau$ is the current concentration of isocyanate (or methyl oleate with ozone), while D_0 and D_τ are the optical densities of gaseous mixture at the system inlet and outlet initially.

These data suggest that the rate constant of the reaction between ozone and TBTU is much greater than those of the reactions of ozone with the $C = C$ bonds in polymers ($k_3 = 14 \times 10^5$ liter/mol s for natural rubber [30]). It should be noted that the second step of the reaction is relatively slow. This seems to be common feature of all antiozonants. Despite the fact that they are capable of reacting with several ozone molecules, the products of the subsequent reaction steps are ineffective because of the slower rate of the reaction with ozone. The introduction into the system of a variety of antiozonants instead of the reference olefin has made it possible to determine, from the depression of SO_2 release, the rate constants of their reactions with ozone. These values together with their rate constants with respect to the ozone dissolution rate are given in Table 14.1.

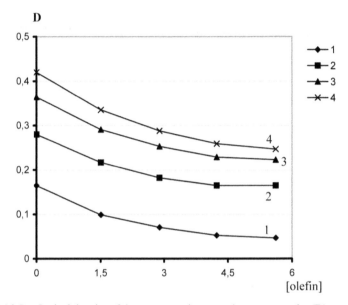

FIGURE 14.5   Optical density of the gaseous mixture at the reactor outlet (D) versus olefin concentration [olefin] at various times: 1, 1 min; 2, 2 min; 3, 3 min; 4, 4 min. Tributylthiourea, $4 \cdot 1 \times 10^{-4}$ mol/liter; $[\text{olefin}]_0 = 1 \cdot 5 \times 10^{-3}$ mol/liter; $V_{O_3} = 120$ ml/min; $1 = 290$ nm

The tabulated data indicate that the protective action is exerted only by those compounds which react with ozone at a rate exceeding the rate of the reaction between ozone and the C = C bonds in macromolecules. The greater the reaction rate constant, the more effective the protection by the compounds tested. Figure 14.6 illustrates the relative efficiency τ_{an}/τ_c expressed as the ratio between the time to cracking in stabilized rubber and the time to cracking in the control specimen, [31] versus the ratio between the rate constants of the ozone-antiozonant and ozone ->C=C< reactions (k_{an}/k_{12}). It can be seen that the resulting curve is linear in semilogarithmic coordinates (15.10).

$$\log {}^{\tau_{an}}\!\big/\!_{\tau_c} = f\left({}^{k_{an}}\!\big/\!_{k_{12}}\right) \tag{14.10}$$

It was established in these experiments that only one class of compounds, namely the thioureas, is excluded from the series. Their experimentally measured activity by this method turnedout to be abnormally high. During tests of tires under normal operating conditions or during dynamic tests this anomaly was not observed. It was natural to assume that the observed deviation is due not the chemical properties of thiourea but to its specific physical behavior in rubbers [18, 22, 32], in particular its tendency to accumulate on the rubber surface to a greater extent than other substances examined. This is also corroborated by published data concentrating the relative content of various antiozonants on the surface when they were initially incorporated in equal amounts [33].

TABLE 14.1 Rate constants of the reaction between antiozonants and ozone and their protective action in rubber [27]

Compound	Formula	Time to Crackig (min)	$k_{21} \times 10^{-6}$ (liter/ mol s)
N,N'-di-n-jctyl-p-Phenyl-ene-diamine	C_8H_{17}—NH—⬡—NH—C_8H_{17}	840	7
N,N'-diisopen-tyl-p-pheny-lenediamine	C_5H_{11}—NH—⬡—NH—C_5H_{11}	870	8
N-Phenyl-N'-isopropy-p-phenylene-diamine	⬡—NH—⬡—NH—$CH(CH_3)_2$	500	7
N,N'-di-a-methylben-zyl-p-phenylene-diamine	⬡—CH(CH₃)—NH—⬡—NH—CH(CH₃)—⬡	250	5
N-a-methyl-benzylanisi-dine	⬡—CH(CH₃)—NH—⬡—OCH_3	90	4

N-butil-N,N'-dibutylthio-urea	C_4H_9NH ——C ——$N(C_4H_9)$; $\overset{\displaystyle \|}{S}$	790	2
Methyl oleate	CH_3O —— C ——$(CH_2)_7$— CH$=$CH——$(CH_2)_7$——CH_3 $\overset{\displaystyle \|}{O}$	80	1

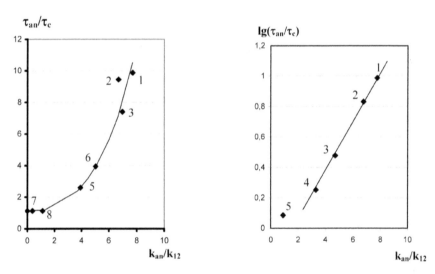

FIGURE 14.6   Relations $\tau_{an}/\tau_c = f(k_{an}/k_{12})$ (a) and $log\,\tau_{an}/\tau_{c=} = f(k_{an}/k_{12})$ (b). 1, N,N'-dioctyl-p-phenylenediamine; 2, N,N'-di-isopentyl-p-phenylenediamine; 3, N-phenyl-N'-isopropyl-p-phenylenediamine; 4, N,N'-di-α-methylbenzyl-p- phenylenediamine; 5, N-α-methyl-benzylanisidine; 6, N,N,N'-tributylthiourea; 7, 2,2-thio-bis-(6-$tret$-butil-4-methylphenol); 8, methyl oleate

The relation $log\ \tau_{an}/\tau_c = f(k_{an}/k_{12})$ is interesting in two respects. Firstly, it can be used as a basis for the development of quantitative methods of evaluating the efficiency of antiozonants. At present no such methods are available; there are only qualitative estimates: good, satisfactory, poor, and so on [32], which make it impossible to compare compounds of different classes or compounds of the same class taken in different amounts. Secondly, it can be an instrument in approximate calculations of maximum antiozonant efficiency. If it is assumed that this relation persists when the rate constant k_{an} increases to a reasonable value (ca 10^8 liter/mol s), even rough estimates show that the efficiency of such an antiozonant will exceed that of the currently available ones by several orders of magnitude.

The mechanism of the physical protection of vulcanizates of unsaturated polymers against ozone is mach less well understood and theoretical interpretations have not yet been elaborated. It is known that the introduction of various waxes into the original stock enhances the ozone resistance of vulcanizates, particularly under static conditions [34–36]. Waxes may be used individually and in combination with antiozonants [18, 37], and it is generally considered that waxes form a protective film on the rubber surface, which slows down the reaction between ozone and the $C = C$ bonds of the polymer. The permeability of the film is associated with the temperature of the wax film, the temperature at which it starts to soften, and plasticity [38].

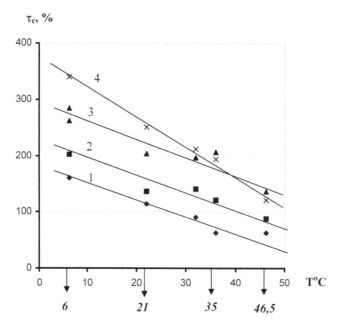

FIGURE 14.7 t_c (expressed as percentage of t_c without wax) for unfilled vulcanizates of natural rubber, containing five parts by weight of paraffin wax, at various test temperatures (°C): 1, 25; 2, 35; 3, 45; 4, 57. e = 10 percent; $[O_3] = 4·5 ' 10^{-7}$ mol/liter.

Figure 14.7 shows the results of determining the time to cracking (τ_c) in the case of unfilled natural rubber containing paraffins with different softening points. The efficiency of waxes is higher at low strains; at high strains, the number of cracks decreases in the presence of waxes but then overstrains ahead of crack and crack propagation rate may even increase.

The chemical composition of waxes influences the protection efficiency. The presence of $C = C$ bonds, amino groups, metal salts and other additives in waxes increases the time to cracking and time to degradation of the specimen [39].

KEYWORDS

- **Antiozonant**
- **Double bonds**
- **Elastomers**
- **Kinetics,**
- **Mechanism**
- **Protective action**
- **Rubbers**

REFERENCES

1. Gorbunov, B. N.; Gurvitch, A.Ya.; Maslova, I. P.; Chimiya i Technologiya stabilizatorov Polymernyh materialov. M., Chimiya, **1981**.
2. Kuzminskii, A. S.; Kavun, S. M.; and Kirpichev, V. P.; Phisiko-chimicheskie osnovy polucheniya, pererabotki I primeneniya elastomerov. M., Chimiya, **1976**.
3. Zuev, Yu.S.; Degteva, T. G.; Stoikost elastomerov v ekspluatatcionnyh usloviyah., M., Chimiya, **1986**, 264 p.
4. Franco Cataldo. Polymer Degradation and Stability, **2001**, *73*, 511–520.
5. Muhutdinov, E. R.; and Diakonov, G. S.; Vestnik Kazanskogo Universiteta, **2010**, *10*, 483–504.
6. Ionova, N. I.; Zemskii, D. N.; Dorofeeva, Yu.N.; Kurliand, S. K.; and Mohnatkina E. G.; *Kautchuk i Rezina*. **2011**, *1*, 9–12.
7. Razumovskii, S. D.; Rakovskii, S. K.; Shopov, D. M.; and Zaikov, G. E.; Ozone and its reactions with organic compounds, Bulgarian Ak.Nauk, Sofia, **1983**, *289* p.
8. Veith, A. G., *Rubber Chem. Technol.* **1972**, *45*, 293.
9. Lipkin, A. M.; Grinberg, A. E.; Gurvich, Ya. A.; Zolotarevskaya, L. K.; Razumovskii, S.; D.; and Zaikov, G. E.; *Vysokomol. Soedin.*, **1972**, *A14*, 78.
10. Lake, G. I.; *Rubber Chem. Technol.*, *43*, **1970**, 1230.
11. Zuev, Yu. S.; and Pravednikov, S. I., *Kauch. Rezina.* **1961**, *1*, 30.
12. Razumovskii, S. D. and Batashova, L. S., *Vysokomol.Soedin.* **1969**, *A9*, 588.
13. Zuev, Yu. S.; Koshelev, F. F.; Otopkova, M. A.; and Mikhaleva, S. B.; *Kauch. Rezina*, **1965**, *8*, 12.
14. Lorenz, O and Parks, C. R.; *Rubb. Chem. Technol.* **1963**, *36*, 194.
15. Layer, R. W.; *Rubber Chem. Technol.* **1966**, *39*, 1584.
16. Razumovskii, S. D.; Buchachenko, A. L.; Shapiro, A. B.; Rozantsev, E. G.; and Zaikov, G. E.; *Dokl. Akad. Nauk SSSR.* **1968**, *183*, 1106.

17. Murray, R. W.; and Story, P. R.; Chemical Reactions of Polymers, Vol. 2, Ed. E. M. Fettes, Wiley-Interscience: New York; **1964**.
18. Zuev, Yu. S.; *Razrushenie Polymerov pod Deistviem Agressivnykh sre*. Khimiya, Moscow. (a) Ibid. p. 103.
19. Lipkin, A. M.; Zolotarevskaya, L.; K.; Grinberg, A. E.; Gurvich, Ya.; A.; Razumovskii, S. D.; and Zaikov, G. E.; *Vysokomol.Soedin*. **1972**, *A14*, 680.
20. Ambelang, J. C.; Kline, R. H.; Lorenz, O. M.; and Parks, C. R.; *Rubber Chem. Technol*. **1963**, *36*, 1497.
21. Tucker, H.; *Rubber. Chem. Technol*. **1959**, *32*, 269.
22. Zuev, Yu. S.; *Zhurn. Vses. Khim. Obschchestvo Mendeleeva*. **1965**, *11*, 288.
23. Gorbunov, B. N.; Gurvitch, A.Ya.; Maslova, I. P., *Ozone; Chemistry and Technology*. A Review of the Literature, 1961–1974; **1976**. Philadelphia.
24. Braden, M.; *J. Appl. Polym. Sci.* **1962**, 86.
25. Delman, A. D., Sims, B. B. and Allison, A. R., *J. Anal. Chem*. **1954**, *26*, 1589.
26. Razumovskii, S. D. and Zaikov, G. E., *Ozone and its Reactions with Organic Compounds* (1974), Nauka, Moscow.
27. Lipkin, A. M., Razumovskii, S. D. Gurvich, Ya. A., Grinberg, A. E., and Zaikov, G. E., *Dokl. Akad. Nauk SSSR*, **1970**, *192*, 127.
28. Parfenov, V. M., Rakovski, S. K., Shopov, D., M., Popov, A. A. and Zaikov, G. E., *Izves. Khim. Bolg. Akad. Nauk*, **1978**, *11*, 180.
29. Rakovsky, S. K., Cherneva, D. R., Shopov, D. M. and Razumovskii, S. D., *Izves. Khim. Bolg. Akad. Nauk*. **1976**, *9*, 711.
30. Kefeli, A. A.; Vinitskaya, E. A.; Markin, V. S.; Razumovskii, S. D.; Gurvich, Ya.A.; Lipkin, A. M.; and Neverov, A. N.; *Vysokomol. Soedin*. **1977**, *A19*, 2633.
31. Lipkin, A. M.; Grinberg, A. E.; Gurvich, Ya.A.; Zolotarevskaya, L. K.; Razumovskii, S. D.; and Zaikov, G. E.; *Vysokomolek.soedin*. **1972**, *A14*, 87.
32. Ambelang, I. C.; and Lorenz, O.; *Rubber. Chem. Technol*. **1963** *36*, 1533.
33. Hogkinson, G. T.; and Kendall, C. E.; In Proceedings of the 5th Rubber Technol. Conference, Ed. T.H.; Messenger, Institution of the Rubber Industry, London, **1962**.
34. Bennet, H.; Commercial Waxes, 2nd edition; Chemical Publishing Co.: New York; **1956, p. 192**.
35. Backer, D. E.; *Rubber Age (NY)*. **1955**, *77*, 58.
36. Buswell, A. G. and Watts, J. T., *Rubber Chem. Technol*. **1962**, *35*, 421.
37. Zinchenko, N. P. and Vinogradova, T. N., *Zashchita Shinnykh Rezin ot Vozdeistviya Ozona I Utomleniya*. TsNIIEneftekhim, Moscow; **1969**.
38. Zuev, Yu. S.; and Karandashev, B. P.; *Plastifikatory I Zashchitnye Agenty iz Neftyanogo Syrya*, Ed. I. P. Lukashevich, . Khimiya, Moscow, **1970**; p. 161.
39. Zuev, Yu. S. and Postovskaya, A. F., *ibid, p. 136.*

CHAPTER 15

SYNTHESIS OF 3'-α-FLUORONUCLEOSIDES USING PYRIMIDINE NUCLEOSIDE PHOSPHORYLASE OF *Thermus thermophilus* AND PURINE NUCLEOSIDE PHOSPHORYLASE OF *ESCHERICHIA COLI*

A. I. BERESNEV[1], S. V. KVACH[1], G. G. SIVETS[2*], and A. I. ZINCHENKO[1*]

[1]Institute of Microbiology, National Academy of Sciences,

220141, Kuprevich Str., 2, Minsk, Belarus.

**E-mail: zinch@mbio.bas-net.by; Fax: +375(17)264-47-66

[2]Institute of Bioorganic Chemistry, National Academy of Sciences,

220141, Kuprevich Str., 5/2, Minsk, Belarus.

*E-mail: gsivets@yahoo.com; Fax: +375 (17) 267-87-61

CONTENTS

15.1   INTRODUCTION

Progress of genetical engineering triggered development of recombinant microbial strains providing for hyperproduction of valuable enzymes—nucleoside phosphorylases involved in enzymatic synthesis of nucleosides. Mesophilic bacteria *Escherichia coli* are most frequently used for construction of new recombinant strains [1, 2]. A series of recent publications described production and properties of pyrimidine and purine nucleoside phosphorylases generated by various thermophilic microorganisms, their activity toward pyrimidine and purine nucleosides, and biotechnological potential to derive purine 2'-fluoronucleosides [3]. Thermal stability of these biocatalysts enables to run enzymatic processes at elevated temperatures resulting in increased substrate solubility, accelerated nucleoside-yielding procedure in the course of enzymatic transglycosylation reaction, and its enhanced efficiency in comparison with chemical methods [4].

Bacteria *E. coli* are widely used for production of recombinant thermophilic pyrimidine and purine nucleoside phosphorylases, though low level of biocatalysts synthesis by bacterial cells should be noted. The main reason for low efficiency of geterologous expression of genes cloned in *E. coli* is formation of mRNA pins at initial transcript site obstructing ribosomal synthesis of polypeptide chain [5]. Earlier we engineered strain *E. coli* KNK-12/1 producing pyrimidine nucleoside phosphorylase (PyrNPase) of *Thermusthermophilus* and originally demonstrated that the resulting thermophilicPyrNPase displayed a unique capacity to catalyze reaction of 3'-fluoro-2',3'-dideoxythymidine (3'-F-2',3'-ddThd) phosphorolysis yielding α-D-pentofuranose-1-phosphate [6].

Enzymatic synthesis of 3'-fluoro-2',3'-dideoxyguanosine (3'-F-2',3'-ddGuo), a potential chemical therapy agent for treatment of viral infections, was accomplished using nucleoside phosphorylases and adenosine deaminase from *E. coli.*

Since PyrNPase of *T. thermophilus* is a thermostable protein, level of its synthesis in *E. coli* cells is relatively low, which limits enzyme application prospects for synthesis of modified purine nucleosides [3].

Previously constructed recombinant strain *E. coli* BM-D6 produced homologous purine nucleoside phosphorylase (PurNPase) acting (like PyrNPase) as a key catalyst transforming pyrimidine nucleosides into modified purine nucleosides via enzymatic transglycosylation reaction [7]. PurNPase catalyzes stereoselective reaction of intermediate α-D-pentofuranose-1-phosphate (product of pyrimidine nucleoside phosphorolysis mediated by PyrNPase) condensation with purine heterocyclic base leading to formation of modified purine nucleoside.

Aim of this research was to increase efficiency of producing *T. thermophilus*PyrNPase in cells of genetically engineered *E. coli* strain by optimizing the structure of the respective translated mRNA and to investigate enzymatic synthesis of purine 3'-fluoro-3'-deoxy- and 3'-fluoro-2',3'-dideoxynucleosides possessing antiviral and cytostatic activities from the available pyrimidine nucleosides engaging tandem reactions in the presence of recombinant nucleoside phosphorylases [8].

15.2 MATERIALS AND METHODS

Strain *E. coli* DH5α (Invitrogen, USA) essential for generation of constructed recombinant vector, strain *E. coli* BL21(DE3) (Invitrogen, USA) used for expression of cloned genes, recombinant strain *E. coli* BM-D6 producing homologous PurNPase, recombinant strain *E. coli* KNK-12/1—a source of *T. thermophilus*PyrNPase, strain *E. coli* pADD3 engineered to produce recombinant homologous adenosine deaminase were applied in this study.

Data of primary structure of gene encoding PyrNPase of *T. thermophilus* were taken from GenBank base of nucleotides sequences. Molecular optimization of mRNA nucleotide sequence related to initial site of transcribed gene composed of 35 nucleotides was carried out using RNAFold appendix of Vienna RNA software package (version 1.8.4). Selection of primers for PCR was based on the available data concerning primary structure of *T. thermophilus*PyrNPase. Restriction sites *Nde*I and *Sal*I linked to 5'-ends of the primers were anchored by 6 random nucleotides. Point mutations in initial segment of PyrNPase gene locus affecting secondary mRNA structure were induced by direct oligonucleotide primer. Reaction mixture for PCR assay (50 µl) included 67 mM Tris-HCl buffer (pH 8.3) comprising 17 mM $(NH_4)_2SO_4$, 3 mM $MgCl_2$, 0,1 percent Tween 20, 0.12 mg/ml BSA, 8 percent glycerol, four 2'-deoxynucleoside triphosphate (each in 0.2 mM concentration), 10 pmol of primers TGTATATCTCCTTCT-TAAAGTTAAACAAAATTATTTC and TATGAACCCCGTAGCATTTATC-CGGGAGAAGCGGGA, 150–200 ng of genomic DNA from *T. thermophilus* plus 1 U of *Pfu*-polymerase. Amplification procedure occurred as follows: 1 min at 98°C, (1 s at 98°C, 10 s at 60°C, 45 s at 72°C)—25 cycles, 2 min at 72°C. PCR products were separated by electrophoresis in 1,5 percent agarose gel. The putative PyrNPase gene product was recovered and ligated into vector pET42a previously digested with *Nde*I and *Sal*I and dephosphorylated with alkaline phosphatase. A novel *E. coli* strain producing heterologous mutant PyrNPase (mutPyrNPase) was derived from this vector.

The bacteria were cultured at temperature 37°C and 42°C on Luria-Bertani medium supplemented with a 50 µg/ml kanamycin. 0.5 mM IPTG was applied to induce synthesis of target enzymes in microbial cells. Process of PyrNPase and mutPyrNPasesynthesis by *E. coli* was monitored by electrophoresis in polyacrylamide gel containing sodium dodecylsulphate (SDS-PAGE). The percent-age of enzymes of total cell proteins was calculated using TotalLab 120 soft-ware.

The grown bacterial cells were harvested by centrifugation, suspended in 10 mM potassium-phosphate buffer, pH 7.0 (PPB) and subjected to ultrasonic dis-integration. Enzyme purification was achieved by heating bacterial sonic lysates comprising PyrNPase and mutPyrNPase from *T. thermophilus* at 80°C during 1 h with subsequent removal of denatured proteins by centrifugation.

PyrNPase and mutPyrNPase activities were evaluated spectrophotometri-cally via changes in absorbance (λ = 300 nm) during phosphorolysis cleavage of thymidine. A unit of pyrNPase and mutPyrNPase activity was defined at the amount of enzyme sufficient to produce 1 µmole of thymine in 1 min under reaction conditions.

Process of 3'-fluoro-3'-deoxyuridine (3'-F-3'-dUrd) phosphorolysis was an-alyzed in the presence of mutPyrNPase: 2.5 U of enzyme activity were added to 3'-F-3'-dUrd in 2 mM PPB and the reaction was performed at 80°C with regular monitoring of phosphorolysis degree in the course of 96 h.

Enzymatic synthesis of purine 3'-fluoro-3'-deoxy- and 2',3'-dideoxy-β-D-ribonucleosides was conducted from 3'-F-3'-dUrd and 3'-F-2',3'-ddThd sub-strates produced by chemical methods described earlier. The reaction were car-ried out in 5 mM PPB at temperature 60°C during 5 days with the following variations in mixture composition: (1)—15 mM 3'-F-3'-dUrd, 10 mM adenine (Ade), 20 U/ml mutPyrNPase *T. thermophilus*, 200 U/ml PurNPase*E. coli*; (2)—15 mM 3'-F-3'-dUrd, 10 mM 2-aminoadenine (2NH$_2$-Ade), 20 U/ml mutPyrN-Pase *T. thermophilus*, 200 U/ml PurNPase*E. coli*; (3)—15 mM 3'-F-3'-dUrd, 10 mM 2NH$_2$-Ade, 20 U/ml mutPyrNPase *T. thermophilus*, (after cooling to room temperature 200 U/mlof adenosine deaminase from *E. coli*was supplied to the reaction mixture and incubated at 25°C for 24 h); (4)—15 mM 3'-F-2',3'-ddThd, 10 mM 2-fluoroadenine (2F-Ade), 20 U/ml mutPyrNPase *T. thermophilus*, 200 U/ml PurNPase*E. coli*; (5)—15 mM 3'-F-2',3'-ddThd, 10 mM 2-chloroadenine (2Cl-Ade), 20 U/ml mutPyrNPase *T. thermophilus*, 200 U/ml PurNPase*E. coli*; (6) – 15 mM 3'-F-3'-dUrd, 10 mM 2F-Ade, 20 U/ml mutPyrNPase *T. thermoph-ilus*, 200 U/ml PurNPase*E. coli*; (7)—15 mM 3'-F-3'-dUrd, 10 mM 2Cl-Ade, 20 U/ml mutPyrNPase *T. thermophilus*, 200 U/ml PurNPase*E. coli*.

Product accumulation was controlled by thin-layer chromatography on Sil-ica gel 60F$_{254}$ plates (Merck, Germany). Chromatographic separation of com-ponents in reaction mixtures № 1, 2, and 3 was performed in the mixture of isopropanol–chloroform–25% aqueous ammonia in the ratio 10:10:1 (v/v), in the reaction mixture № 4 and 5—with chloroform–ethanol in 4:1 ratio (v/v), in reaction mixtures № 6 and 7—with n-butanole–25 percent aqueous ammonia in 7 : 1 ratio (v/v).

Location of substrates and products on the TLC-plate were registered in UV-light and the compounds were eluted in 10 mM PPB. Concentration of products in eluates was determined spectrophotometrically using standard coefficients of molar extinction. The absorbance spectra were recorded at spectrophotometer UV 1202 (Shimadzu, Japan).

The structure of newly synthesized nucleosides was confirmed by comparing their TLC and UV-spectroscopy data with reference specimens of purine-3'-fluoronucleosides obtained earlier by chemical methods and in some cases by ^{1}H NMR, UV, and mass spectroscopy analyses after chromatography on silica gel plates.

The provided experiments values are confidence level of arithmetical means at 95 percent confidence Interval.

15.3   RESULTS AND DISCUSSION

It was stated above that one of the methods of mRNA structural optimization leading to increased efficiency of PyrNPase production by mesophilic bacteria *E. coli* is insertion of silent mutations into initial gene fragment encoding this enzyme. As a result it raises Gibbs free energy (G) of mRNA from modified gene locus and consequently reduces the chance of pin emergence. Using RNAFold application of Vienna RNA software package we found that for mRNA gene domain lacking silent mutations G equaled—1.6 kJ (Figure 15.1a). In this study 3 nucleotide substitutions spanning first 35 base pairs of PyrNPase gene locus enabled to increase G value to—5.9 kJ (Figure 15.1b).

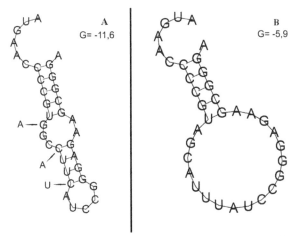

FIGURE 15.1 mRNA secondary structure of initial fragment of transcribed *T. thermophilus*PyrNPase gene prior to (a) and after (b) insertion of mutation

The level of PyrNPase and mutPyrNPase synthesis by *E. coli* cells was compared at standart (37°C) and elevated temperature (42°C). Results of experiments are illustrated by Figure 15.2. Computer processing of electrophoresis data demonstrated that the amount of synthesized bacterial mutPyrNPase exceeded that of PyrNPase by 1.7 times.

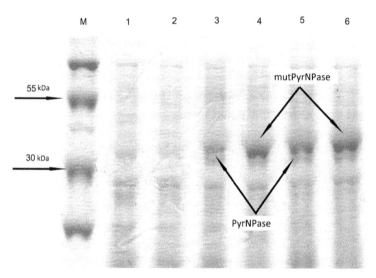

FIGURE 15.2 IPTG induction of synthesis in recombinant *E. coli* cells of PyrNPase and mutPyrNPase from *T. thermophilus* at various temperatures.
M—marker proteins; 1—lisate of noninduced cells producing PyrNPase; 2 – lisate of noninduced cells producing mutPyrNPase; 3—lisate of induced cells producing PyrNPase at 37°C; 4—lisate of induced cells producing mutPyrNPase at 37°C; 5—lisate of induced cells producing PyrNPase at 42°C; 6—lisate of induced cells producing mutPyrNPase at 42°C;

The information collected with the aid of TotalLab 120 was corroborated by measurements of enzyme activities of target proteins. The data are presented in Table 15.1. It was established that induction of enzyme synthesis by *E. coli* cells at 37°C and 42°C resulted in enhanced (1.7 times) mutPyrNPase activity expressed in U/g cells as compared to PyrNPase. At the next stage of research reaction of 3'-F-3'-dUrd phosphorolysis mediated by mutPyrNPase was examined. Experimental results are reflected in Figure 15.3. It was revealed that in the course of 4 day reaction engaging mutPyrNPasephosphorolysis degree of 3'-F-3'-dUrd reached 48 percent allowing to estimate it as a donor in tandem reaction of enzymatic transglycosylation yielding hardly available purine-3'-fluoronucleosides.

TABLE 15.1 Effect of temperature on induction of synthesis of T. thermophilusPyrNPase and mutPyrNPase inE. coli cells

Induction temperature, °C	PyrNPase		mutPyrNPase	
	U/mgcell mass	U/Lcultural liquid	U/mgcell mass	U/Lcultural liquid
37	4,5±0,2	7 100±50	7,5±0,5	12,7±0,9
42	8,2±0,6	11 800±130	12,7±0,9	18 450±220

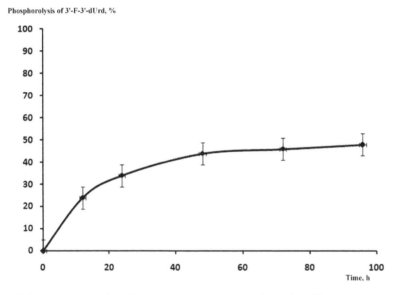

FIGURE 15.3 Dynamics of 3'-F-3'-dUrd phosphorolysis catalyzed by mutPyrNPase from *T. thermophilus*

Enzymatic synthesis of novel and well-known purine 3'-fluoronucleosides previously produced exclusively by chemical methods was carried out from pyrimidine nucleosides (3'-F-3'-dUrd and 3'-F-2',3'-ddThd) and purine heterocyclic bases (2Cl-Ade, 2F-Ade, 2NH$_2$-Ade, Ade) in the presence of recombinant pyrimidine—and purine nucleoside phosphorylases as biocatalysts of phosphorolysis and stereoselective heterobase glycosylation with α-D-pentofuranose-1-phosphate, respectively. Completed reactions resulted in generation of modified purine nucleosides, like 3'-fluoro-3'-deoxyadenosine (3'-F-3'-dAdo), 3'-fluoro-3'-deoxy-2-aminoadenosine (3'-F-3'-d-2NH$_2$-Ado), 3'-fluoro-3'-deoxyguanosine (3'-F-3'-dGuo), 3'-fluoro-2',3'-dideoxy-2-fluoroadenosine (3'-F-2',3'-dd-2F-Ado), 3'-fluoro-2',3'-dideoxy-2-chloroadenosine (3'-F-2',3'-dd-2Cl-Ado), 3'-fluoro-3'-deoxy-2-fluoroadenosine (3'-F-3'-d-2F-Ado), 3'-fluoro-3'-deoxy-2-chloroadenosine (3'-F-2',3'-d-2Cl-Ado). The yield of end products (mol %) was calculated as the percentage of purine nitrogen bases fed into initial reaction mixture. Components of studied enzymatic reactions and yields of target nucleosides are provided in Table 16.2.

TABLE 15.2   Synthesis of fluorinated nucleosides

Number of Reaction Mixture	Product	Acceptor (10 mM)	Donor (15 mM)	Time, h	Product yield, mol.%
1	3'-F-3'-dAdo	Ade	3'-F-3'-dUrd	72	55±2
2	3'-F-3'-d-2NH$_2$-Ado*	2,6-DAP	3'-F-3'-dUrd	72	63±3
4	3'-F-2',3'-dd-2F-Ado	2F-Ade	3'-F-2',3'-ddThd	96	64±2
5	3'-F-2',3'-dd-2Cl-Ado	2Cl-Ade	3'-F-2',3'-ddThd	96	62±3
6	3'-F-3'-d-2F-Ado	2F-Ade	3'-F-3'-dUrd	96	53±2
7	3'-F-3'-d-2Cl-Ado	2Cl-Ade	3'-F-3'-dUrd	96	52±2

*3'-F-3'-d-2NH2-Ado was transformed into 3'-F-3'-dGuoby introducing recombinant adenosine deaminase of E. coli into reaction mixture.

Summing up the obtained results, it should be noted that biotechnological synthesis of purine 3'-fluorinated nucleosides from corresponding pyrimidine analogs catalyzed by recombinant nucleoside phosphorylases appears extremely attractive in terms of producing new purine nucleoside derivatives and further elaboration of chemical-enzymatic processes generating biologically active compounds of this category. The proposed approach in some aspects is superior in efficiency than conventional chemical methods of producing similar modified purine nucleosides based, as a rule, on multistage scheme leading to output of anomeric nucleotide mixtures.

15.4   CONCLUSION

Using advanced genetic engineering methodology, a series of silent mutations was originally introduced into initial fragment of gene cloned in *E. coli* and coding for PyrNPase of *T. thermophilus*. As a result 70% rise in the synthesis level of studied thermostable enzyme was achieved. We were to establish that pyrimidine fluorodeoxynucleosides—3'-F-3'-dUrd and 3'-F-2',3'-ddThd may serve as substrates of phosphorolysis reaction catalyzed by thermostablePyrNPase of *T. thermophilus*. Moreover, recombinant nucleoside phosphorylases were never before involved in enzymatic synthesis of biologically significant purine 3'-fluoronucleosides, such as 3'-F-3'-dAdo, 3'-F-3'-d-2NH$_2$-Ado, 3'-F-3'-dGuo, 3'-F-2',3'-dd-2F-Ado, 3'-F-2',3'-dd-2Cl-Ado, 3'-F-3'-d-2F-Ado and 3'-F-2',3'-d-2Cl-Ado.

KEYWORD

- **Modified nucleosides**
- **Purine nucleoside phosphorylase**
- **Silent mutations**
- **Pyrimidine nucleoside phosphorylase**

REFERENCES

1. Mikhailopulo, I. A.; and Miroshnikov, A. I.; *ActaNatur.* **2010**, *2*(2), 36.
2. Mikhailopulo, I. A.; and Miroshnikov, A. I.; *Mendeleev. Commun.* **2011**. *21*, 57.
3. Szeker, K.; Niemitalo, O.; Casteleijn, M. G.; Juffer, A. H.; Neubauer, P.; *J. Biotechnol.* **2011**. *156*, 268.
4. Szeker, K.; Zhou, X.; Schwab, T.; Casanueva, A.; Cowan, D.; MIKhailopulo, I.A.; Neubauer, P.; *J. Mol. Catal. B: Enzym.* **2012**. 84, 27.
5. Zhu, S.;, Ren, L.; Wang, J.; Zheng, G.; and Tang, P.; *Bioorg. Med. Chem. Lett.* **2012**, *22*(5), 2102.
6. Erasneu, A.; Kvach, I. B, S. V.; Sivets, G. G.; and Zinchenko, A. I.; Proceedings of the National Academy of Sciences of Belarus. Series of Biological Sciences. **2013**, *4*, 71 (in Russian).
7. *Burko*, D. V.; Eroshevskaya, L. A.; Kvach, S. V.; Shakhbazau, N.A. Kartel, and A. I. Zinchenko; Biotechnology in Medicine, Foodstuffs, Biocatalysis, Environment and Biogeotechnology. Nova Science Publishers, Inc, New York, **2010**; 1 p.
8. Qui, X. I.; Xu, X. H.; and Qing, F. L.;*Tetrahedron.* **2010**, *66*, 789.

PART III
BIOLOGICAL SCIENCE

CHAPTER 16

APPLICATION STABLE RADICALS FOR STUDY OF BEHAVIOR OF BIOLOGICAL SYSTEMS

M. D. GOLDFEIN and E. G. ROZANTSEV

Saratov State University named after N.G. Chernyshevsky, Russia
E-mail: goldfeinmd@mail.ru

CONTENTS

16.1   INTRODUCTION

The presence of paramagnetic particles in liquid or solid objects opens new opportunities of their studying by the EPR technique. Ready free radicals and substances forming paramagnetic solutions due to spontaneous homolization of their molecules in liquid and solid media (i.e., triphenylmethyl dimer, Frémy's salt, or 4,8-diazaadamantan-4,8-dioxide) can act as sources of paramagnetic particles.

The experimental technique of radio spectroscopic examination of condensed phases with the aid of paramagnetic impurities is usually called the paramagnetic probe method. Though iminoxyl radicals have found broadest applications for probing of biomolecules, nevertheless, the first application of the paramagnetic probe technique to study a biological system is associated with a quite unstable aminazine radical cation:

The progress in the theory and practice of EPR usage in biological research is restrained by the narrow framework of chemical reactivity of nonfunctionalized stable radicals with a localized paramagnetic center like

The substances of this class only enter into common, well-known free radical reactions, namely: recombination, disproportionation, addition to multiple bonds, isomerization, and β-splitting [1]. All these reactions proceed with the indispensable participation of a radical center and steadily lead to full paramagnetism loss. And still the synthesis of nonfunctionalized stable radicals plays a very important role. No expressed delocalization of an uncoupled electron over a multiple bond system has been shown to be obligatory for a paramagnetic to be stable.

Despite of the basic importance of the discovery of stable radicals of a non-aromatic type [2], this event has not changed contemporary ideas on the reactivity of stable radicals.

In the early 1960s, one of the authors of this book laid the foundation of a new lead in the chemistry of free radicals, namely, the synthesis and reactivity of functionalized stable radicals with an expressed localized paramagnetic center [3].

The opportunity to obtain and study a wide range of such compounds with various functional substituents arose in connection with the discovery of free radical reactions with their paramagnetic center unaffected.

Functionalized free radicals have found broad applications as paramagnetic probes for exploring molecular motion in condensed phases of various natures. The introduction of a spin label technique (covalently bound paramagnetic probe) is associated with their usage; its idea is not new and based on the dependence of the EPR spectrum shape of the free radical on the properties of its immediate atomic environment and the way of interaction of the paramagnetic fragment with the medium. The reactions of free radicals with their paramagnetic center unaffected (Neumann–Rozantsev's reactions) [4] have become the chemical basis of obtaining spin-labeled compounds.

The concept of the usage of nonradical reactions of radicals to study macromolecules was formulated at the Institute of Chemical Physics (USSR Academy of Sciences) by G. I. Lichtenstein [5] in 1961, and the theoretical bases of this method, calculation algorithms for the correlation times of rotary mobility of a paramagnetic particle from their EPR spectrum shape were developed by McConnell [6], Freed and Fraenkel [7], Kivelson [8], and Stryukov [9]. Have investigated the behavior of iminoxyl radicals in various systems and obtained important and interesting results [10].

Let us cite several important aspects of the application of organic paramagnets to researching biological systems.

16.1.1 APPLICATION OF IMINOXYL FREE RADICALS FOR STUDYING OF IMMUNE GAMMA GLOBULINS

From the physicochemical viewpoint, the mechanism of various immunological reactions is determined by changes in the phase state of a system. Despite of the wide use of these reactions in medical practice, the nature of interaction of antigens with antibodies is not quite clear. To study this process, gamma globulins labeled with iminoxyl radicals were used [11]. The molecule of gamma globulin is known to consist of four polypeptide chains bound with each other with disulfide bridges. When the interchain disulfide bonds are split, the polypeptide chains continue to keep together. A spin label (an iminoxyl derivative of maleimide)

was attached to the sulfhydryl groups obtained by restoration of disulfide bonds with β-mercaptoethanol. Experiments were made on the rabbit and human gamma globulins. The EPR spectra in both cases corresponded to rather high mobility of free radicals (correlation times $\tau = 1.1 \cdot 10^{-9}$ s for human gamma globulin and $7.43 \cdot 10^{-9}$ s for rabbit's one). By comparing the correlation times in these proteins with the values obtained in experiments with serum albumin labeled with sulfhydryl groups, treated with urea ($\tau = 1.09 \cdot 10^{-9}$ s) and dioxane ($2.04 \cdot 10^{-9}$ s), it is possible to conclude that the fragments of polypeptide chains bearing free radicals possess no ordered secondary structure. This is in accord with data on very low contents of α-helical structures in gamma globulins. Such a character of the EPR spectrum of immune gamma globulin with preservation of its specific activity opens the possibility to explore conformational and phase transitions at specific antigen–antibody reactions. Sharp distinctions in the mobility of spin labels were revealed at sedimentation of the rabbit antibodies by salting-out with ammonium sulfate and precipitation with a specific antigen (egg albumin). Precipitation of antibodies with a specific antigen led only to a small reduction of the paramagnetic label mobility whereas sedimentation with ammonium sulfate caused strong retardation of the rotary mobility of free radicals (Figure 16.1).

These results can be considered as direct confirmation of the alternative theory [12] according to which the precipitate formation is associated with the immunological polyvalency of the antigen and antibody relative to each other. Really, excepting the location of spin labels inside the antibody's active center, it is possible to conclude that the rather mobile condition of spin labels in the antigen–antibody precipitate may remain if only there is no strong dehydration of the antibodies due to intermolecular interactions. Unlike gamma globulin precipitated with ammonium sulfate, the specific precipitate, according to the lattice theory, has a microcellular structure. At long storage of the antigen–antibody precipitate with no addition of stabilizers, the mobility degree of iminoxyl radicals sharply decreased. This is apparently a result of secondary dehydration of the antibodies owing to protein molecule interactions in the precipitate.

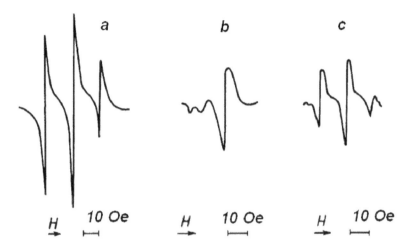

FIGURE 16.1   EPR spectra of gamma globulin labeled with an iminoxyl radical with its free valency unaffected: (**a**) in solution; (**b**) in the precipitate obtained by salting-out with ammonium sulfate; (**c**) in the specific precipitate

16.1.2   EXPLORING STRUCTURAL TRANSITIONS IN BIOLOGICAL MEMBRANES

Biological membranes, in particular, mitochondrial membranes, are known to play a huge role in redox processes in the cell, being the very place of respiratory chain enzyme localization. Baum and Riske [13] have found essential distinctions in the properties of one of the mitochondrial membrane fragments (Complex III of Electron Transport Chain) upon transition from the oxidized form to the reduced one: the sulfhydryl groups, easily titrated in the oxidized form of this fragment, become inaccessible in the reduced form. The nature of trypsin digestion of this complex also strongly changes.

Earlier, changes in the repeating structural units of mitochondria in conditions leading to the formation of macroergic intermediate products or their provision, e.g., active ion transfer, were revealed by electron microscopy [14]. All this has allowed us to assume that any redox reaction catalyzed by an enzymatic chain of electron transfer is accompanied by some kind of "conformational wave" probably covering not only the protein component of the membrane but also a higher level of its organization, namely: fragments of the membrane of more or less complexity degree, including its lipidic part. To verify this assumption, a modification of the spin label method (a method of noncovalent bound paramagnetic probe) was used. The radical is kept by the matrix (membrane) involving only weak hydrophobic bonds. Such an approach allows studying of weak interactions in the system without essential disturbance of the biochemical functions of the biomembrane and its structure. The paramagnetic probe was 2,2,6,6-tetramethylpiperidine-1-oxyl caprylic ester:

This compound was prepared from caprylic acid chloranhydride and 2,2,6,6-tetramethyl-4-oxpiperidin-1-oxyl in a triethylamine medium by a radical reaction with free valency unaffected. The paramagnetic probe was introduced into a suspension of electron transport particles (ETP) isolated from the bull heart mitochondria by the technique described in Ref. at the Laboratory of Bioorganic Chemistry, Moscow State University. These fragments of the mitochondrial membrane are characterized by a rather full set of enzymes of the respiratory chain with the same molar ratio as in the intact mitochondria [14]. The ability of oxidizing phosphorylation, however, is lost under the used way of isolation.

The paramagnetic probe is insoluble in water but solubilized by ETP suspended in a buffer solution. Owing to this, the observed EPR spectrum is free of any background due to the radicals not attached with the object under study. The presence of a voluminous hydrocarbonic chain provides "embedding" of a molecule of the probe into the lipidic part of ETP. Therefore, the EPR spectrum reflects the condition of exactly this fraction of the membrane. To detect conformational transitions, EPR spectra were recorded before and after the introduction of oxidation substrata (succinate and NAD-N), and after oxidation of the earlier reduced respiratory chain with potassium ferricyanide. Typical results are shown in Figure 16.2 [14].

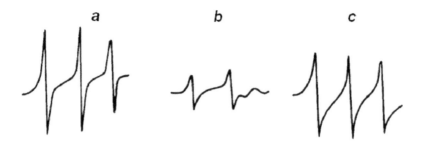

FIGURE 16.2 Changes in the EPR spectrum anisotropy of the hydrophobic iminoxyl radical in a suspension of electron transport particles from bovine heart mitochondria with oxidation substrates added.

The enhanced anisotropy of the EPR spectrum of iminoxyl after substratum introduction is clearly seen. The spectrum in the ferricyanide-oxidized ETP almost does not differ from that in the intact ETP.

ETP inactivation by long storage at room temperature or cyanide inhibition eliminated this effect. Comparison of the shape of signals and correlation times shows that the EPR spectrum of iminoxyl in the intact ETP consists of two signals differing by anisotropy. The radical localized in that part of the membrane where the effective free volume available for radical motion is rather large gives a weakly anisotropic signal. The strongly anisotropic (retarded) spectrum belongs to the radicals localized in other sites of the system with a smaller effective free volume. Reduction of the respiratory chain with substrata leads, owing to cooperative-type conformational transitions, to a reduced fraction of sites with large free volume (i.e. to an increased microviscosity of the immediate environment of the radical). The correlation time of the whole spectrum changes from $20 \cdot 10^{-10}$ s in the oxidized ETP to $4 \cdot 10^{-10}$ s in the reduced ETP.

Concurrently with enhancing the anisotropy of the signal, its intensity decreases as well: iminoxyl reduces, apparently, to hydroxylamine derivatives. Potassium ferricyanide inverts this process. It is necessary to consider that oxidation substrata, themselves, do not interact considerably with iminoxyls. Obviously, the conformational transition not only leads to a changed microviscosity but eliminates any steric obstacles complicating reduction of the radical. In principle, this circumstance points to possibly a new, actually chemical, aspect of application of the paramagnetic probe technique.

16.1.3   EXPLORING THE STRUCTURE OF SOME MODEL SYSTEMS

The application of the paramagnetic probe technique in systems like biological membranes poses a number of questions concerning the behavior of hydrophobic labels in media with an ordered arrangement of hydrophobic chains. As a first stage, mixtures of a nonionic detergent (Tween 80) and water were studied. Tween 80 originates from polyethoxylated sorbitan and oleic acid (polyoxyethylene (20) sorbitan monooleate) and classifies as a nonionic detergent based on polyethylene oxide [14]. Our choice of this object was determined by some methodical conveniences, and also some literature data on the structure of aqueous solutions of Tween, obtained by classical methods (viscometry, refractometry, etc.) The properties of this detergent are also interesting in themselves since it finds quite broad applications for biological membrane fragmentation.

The esters of 2,2,6,6-tetramethyl-4-oxypiperidin-1-oxyl and saturated acids of the normal structure with a hydrocarbonic chain length of 4, 7, or 17 carbon atoms or the corresponding amides were used as a paramagnetic probe. For comparison, the behavior of hydrophobic labels IV and V in these systems was also studied

I

II

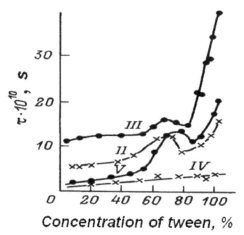

The course of changes in the correlation time of radical rotation as a function of the Tween concentration is shown in Figure 17.3.

FIGURE 16.3 Changes of the correlation time in the detergent–water system, for nitroxyl radicals with a strongly localized paramagnetic center.

It is possible to resolve several areas, apparently, corresponding to various structure types. The initial fragment, only distinguishable for the easiest radicals, corresponds to an unsaturated Tween solution in water. This region is better revealed for the detergents with a higher critical micelle concentration (CMC) (e.g., for sodium dodecyl sulfate) (Figure 16.4) [15, 16]. Then micelle formation occurs. The correlation time of water-insoluble labels increases and comes to a plateau; at further increasing the detergent concentration, it passes through a maximum, and then monotonously increases up to its value in pure Tween. It is useful to compare these data with the results of viscosity measurements by usual macromethods. Figure 16.5 shows that viscosity has one extremum about 60 percent of Tween. Therefore, at a high Tween concentration, the effective volume available for probe molecule rotation and macroviscosity do not correlate.

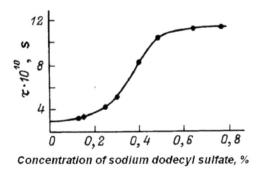

FIGURE 16.4   Estimation of the critical micelle concentration of sodium dodecylsulfate with the aid of 2,2,6,6-tertamethyl-4-hydroxypiperidyl-1-oxyl varerate.

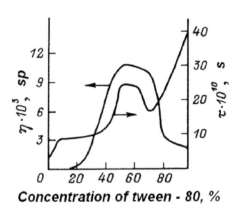

FIGURE 16.5   Changes in the microviscosity and macroviscosity of the water–Tween 80 system.

These results suggest the following interpretation. In pure Tween the lamellar structure provides easy layer-by-layer sliding, hence, the macroviscosity of the system is low. However, in the absence of water, the interaction of the polar groups is strong, the hydrocarbonic chains are ordered, and the effective free volume in the field of radical localization is small. Small water amounts lead to the formation of defects in the layered structure. Sliding is hindered, and the viscosity increases. However, moistening breaks the close interaction of the polar groups in Tween. These groups are deformed, at the same time, the hydrocarbonic chains are disordered. The microviscosity of the hydrocarbonic layer so decreases. Upon termination of hydration of the polar groups, water-filled cavities are formed. They are a structural element (micelle) of which the system is built, for example, a hexagonal P-lattice is formed, by Luzzatti. Structure formation manifests itself as increasing microviscosity and macroviscosity. In the field of the maximum, phase inversion is possible. Structural units of a new type (Tween micelles in water) are formed, passing into colloidal solution upon further dilution. The course of microviscosity changes at high Tween concentrations amazingly resembles the change in correlation time when some lyophilized cellular organelles are moistened. This similarity confirms that in the field of τ maximum, where restoration of the biochemical activity of chloroplasts begins, a phase transition occurs of the same type as in the LC "detergent–water" systems.

Some information on the behavior of radical particles in colloidal systems is provided by the results of temperature measurements. Figure 17.6 illustrates the temperature dependence of the correlation time in Arrhenius' coordinates for several iminoxyl radicals of various hydrophobicity degrees.

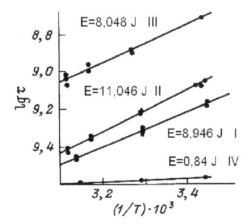

FIGURE 16.6 Activatyion energy of rotational diffusion of free iminoxyl (nitroxyl) radicals in the Tween–water (30% of Tween 80) system.

The strong dependence of the preexponential factor on chain length, apparently, confirms that the observed correlation time really reflects rotation of the radical molecule as a whole. However, this result also allows another interpretation, namely: depending on the hydrocarbonic chain length, the radical introduces into a detergent micelle more or less deeply. This may lead to a changed rotation frequency of iminoxyl groups round the ordinary bonds in the molecule, depending on the environment of the polar end of the radical. If the polar group of the radical is on the micelle–solvent interface, the measured frequency should depend on the surface charge (potential) of the micelle. In our opinion, this circle of colloid chemical problems, closely connected with questions of transmembrane transfer in biological systems, will provide one more application field of the paramagnetic probe technique.

16.1.4 SOLVING OTHER BIOLOGICAL PROBLEMS WITH STABLE RADICALS

Further progress in the field of the usage of stable paramagnets to solve various biological problems is reflected in numerous reviews and monographs [17–20].

For dynamic biochemistry, of undoubted interest are local conformational changes of protein molecules in solutions. The distances between certain loci of biomacromolecules, in principle, can be estimated quantitatively by means of stable paramagnets. Upon introduction of iminoxyl fragments into certain sites of native protein (NRR-method) the distance between the neighboring paramagnetic centers can be calculated from the efficiency of their dipole–dipole interaction in vitrified solutions of a spin-labeled preparation (the EPR method).

The first attempt to estimate the distances between paramagnetic centers in spin-labeled mesozyme and hemoglobin was undertaken by Liechtenstein [21], who indicated prospects of such an approach. In this regard, there appeared a need of identification of a simple empirical parameter in EPR spectra for quantitative assessment of the dipole–dipole interaction of paramagnetic centers.

A convenient empirical parameter was found when studying vitrified solutions of iminoxyl radicals. It was the ratio of the total intensity of the extreme components of a spectrum to the intensity of the central one (Figure 16.7). To establish a correlation of the d_1/d value with the average distance between the localized paramagnetic centers, the corresponding calibration plots are drawn (Figure 16.8). Calculations have shown that the d_1/d parameter depends on the value of dipole–dipole broadening, being in fair agreement with independently obtained experimental results.

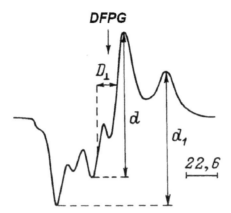

FIGURE 16.7 EPR spectrum of the iminoxyl free biradical (2,2,6,6-tetramethyl-4-hydroxypiperidine-1-oxyl phthalate) vitrified in toluene at 77 K.

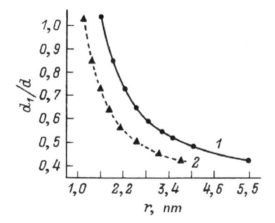

FIGURE 16.8 Dependence of the d_1/d paremeter of an EPR spectrum on the mean distance r between interaction paramagnetic centers of iminoxyl radicals (1) and biradicals (2) at 77 K

Subsequently, the methods of quantitative assessment of the distances between paramagnetic centers in biradicals and spin-labeled biomolecules became reliable tools for structural research.

When determining relaxation rate constants of various paramagnetic centers in solution, the values of these constants have appeared to significantly depend on the chemical nature of the functional groups in iminoxyl radicals.

Sign inversion of the electrostatic charge of the substituent and its distance from the paramagnetic iminoxyl group have the strongest impact on the values of the constants. The electrostatic effect caused by the value and sign of the charge of paramagnetic particles interacting in solution is a more essential factor. An increased ionic strength leads to a change in the k value in qualitative agreement with Debye's theory. The substituent's mass is another factor considerably influencing the value of the constant. If the substituent is a protein macromolecule, the value of k decreases twice. The possibilities of application of the paramagnetic probe technique for detection of anion and cation groups and estimation of distances to explore the microstructure of protein were analyzed. The dependences obtained in experiment show that Debye's equation with $D = 80$ can be successfully applied to experimental data analysis, and, in particular, to estimating the distance from the iminoxyl group of a spin label on protein to the nearest charged group if this distance does not exceed 1.0–1.2 nm.

Considering the role of free radical processes in radiation cancer therapy, it was offered to investigate the influence of iminoxyl radicals upon the organism of laboratory animals. Pharmacokinetic studies [22] have shown that the elementary functionalized derivatives of 2,2,6,6-tetramethylpiperidin-1-oxyl possess rather low toxicity and show an expressed antileukemic activity. The highest values of retardation coefficients were observed for hematopoietic stem cells in peripheral blood and marrow. This circumstance stimulated further structural and synthetic research directed to obtaining more effective and less toxic cancerolytics and sensibilizers for radiation cancer therapy.

After the first publication, for a rather short time, many potential cancerolytics were synthesized in the Laboratory of Stable Radicals, Institute of Chemical Physics, USSR Academy of Sciences, among which of greatest interest from biologists were so-called paramagnetic analogs of some known antitumor preparations, e.g., a paramagnetic analog of Thiotepa [22]:

A brief review on the usage of stable paramagnets of the iminoxyl series in tumor chemotherapy is presented by Suskina [23].

KEYWORDS

- **Biology**
- **Organic paramagnetic method**
- **Radio spectroscopy**
- **Sensing, mechanism.**
- **Study**

REFERENCE

1. Rozantsev, E. G.; *Chem. Encyclopedic.* Dictionary. M. SE. **1983,** p 489 p.[Russ].
2. Rozantsev, E. G.; Lebedev, O. L.; and Kazarnovskii, S. N.; Diploma for the opening number 248 of 05.10.1983. TS. **1982.** No 6.p. 6. [Russ].
3. Rozantsev, E. G.; Doctor dissertation. Moscow: Institute of Chemical Physics, USSR Academy of Sciences, **1965.** [Russ].
4. Zhdanov, R. I.; Nitroxyl Radicals and Non-Radical Reactions of Free Radicals / / Bioactive Spin labels. Zhdanov R. I. Ed Springer: Berlin. Heidelberg. N-Y. 1991. pp. 24.
5. Liechtenstein, G. I.; Method of spin labels in molecular biology, M: Nauka. **1974.** pp. 255. [Russ].
6. McConnell, N. J.; *Chem. Phys.* **1956,** *25,* 709.
7. Freed J.; and Fraenkel, G. J.; *Chem. Phys.* **1963,** *39,* 326.
8. Kivelson, D. *J. Chem. Phys.* **1960.** *33,* 4094.
9. Stryukov, V. B.; Stable radicals in chemical physics. M.: Znanie, **1971.** [Russ].
10. Goldfeld, M. G.; Grigoryan, G. L.; Rozantsev, E. G.; *Polymer* (Gath. of preprints . M. ICP AS SSSR. **1979**.p. 269. [Russ].
11. Grigoryan, G. L.; Tatarinov, S. G.; Cullberg, L. Y.; Kalmanson, A. E.; Rozantsev, E. G.; and Suskina, V. I.; Abst. of USSR Academy of Sciences. **1968,** *178. 31,* pp. 230, pp. 768. [Russ].
12. Pressman, D.; Mol. Structure and Biological Specificity. Washington. D. C. **1957.**
13. Baum, H.; Riske J.; Silman H.; and Lipton, S.; *Froc. Nat. Acad. Sci.* US. **1967,** *57,* 798.
14. Rozantsev, E. G.; Biochemistry of meat and meat products (General Part) . Ed. Manual for students. M. Depi. print. **2006.** p 240. [Russ].
15. Goldfeld, M. G.; Koltover, V. K.; Rozantsev, E. G.; Suskina, V. I.; Kolloid.Zs. **1970.**
16. Schenfeld, H.; Nonionic detergents, **1963.**
17. Smith J.; Shrier–Muchillo Sh.; and Murch, D.; Method of spin labels. Free radicals in biology. M: Mir. **1979** *1,* 179.
18. Method of spin labels and probes, problems and prospects. M: Nauka. **1986.** [Russ].
19. Nitroxyl radicals. Synthesis, chemistry, applications. M: Nauka. **1987.** [Russ].
20. Rozantsev, E. G.; Goldfein, M. D.; Pulin, V. F.; Organic paramagnetic . Saratov: Saratov Govt. University **2000,** 340. [Russ].
21. Liechtenstein, G. I.; *J. Mol. Biol.* **1968,** *2,* 234.

22. Konovalova, N. P.; Bogdanov, G. N.; and Miller, V. B. Abst. of USSR Academy of Sciences. T. **1964**. *157*, 707. [Russ].
23. Suskina, V. I.; Candidate dissertation. M: Academy of Sciences ICP USSR. 1970. [Russ].

CHAPTER 17

BIOPHARMACEUTICALS: AN INTRODUCTION TO BIOTECHNOLOGICAL ISSUES AND PRACTICES

SANJAY KUMAR BHARTI and DEBARSHI KAR MAHAPATRA

School of Pharmaceutical Sciences, Guru Ghasidas Vishwavidyalaya (A Central University), Bilaspur, Chattisgarh, India

CONTENTS

17.1 BIOPHARMACEUTICALS AND BIOSIMILARS

Biopharmaceutical agents are medicinal products of biotechnological origin, which contain proteins derived from recombinant DNA technology and hybridoma techniques, and have revolutionized the treatment of many diseases, including anemia, diabetes, cancer, hepatitis and multiple sclerosis, etc. [1]. Recombinant proteins are derived from cell lines that are maintained in long-term culture, including some that are derived from genetically engineered bacteria (e.g., Escherichia coli). Examples of biopharmaceuticals include biological proteins (e.g., cytokines, hormones and clotting factors), monoclonal antibodies (mAbs), vaccines and cell/tissue-based therapies (refer Table 17.1 for classification) [2].The use of these agents has increased dramatically in recent years. Biological medical products that are biologically similar to registered innovator products are referred to as "biosimilars" in Europe and South-East Asia and 'follow on biologicals' in the USA [3]. Biosimilars are defined as biological products similar, but not identical, to reference products that are submitted for separate marketing approval following patent expiration of the reference products [1]. Table 17.2 shows standard definitions for conventional generic agents, biopharmaceuticals and biosimilars based on terminology used by the European Medicines Agency (EMEA) [4].

TABLE 17.1   Classes of approved recombinant-protein Biosimilars

CLASS	EXAMPLES
Hormones	Insulin (e.g., Humulin), glucagon (e.g., Glucagen), human growth hormone (e.g., Humatrope), thyrotropin (Thyrogen), follicle-stimulating hormone (Gonal-F), lutelnizing hormone (lutropin alfa [Luveris]), human chorionic gonadotropin (Ovidrel), erythropoietin (e.g., epoctin alfa [Epogen])
Cytokines	Interferon alfa (e.g., Roferon-A), granulcyte-colony-stimulating factor (filgrastim [Neupogen]), interleukin (e.g., aldesleukin [Proleukin]) Clotting factors Factor VII (NovoSeven), factor VIII (e.g., Kogenate), factor IX (BeneFIX)
Monoclonal antibodies	Antibodies to vascular endothelial growth factor (bevacizumab [Avastin]), epidermal growth factor receptor (cetuximab [Erbitux]), GPIIb/IIIa receptor (abciximab [ReoPro]), CD20 (rituximab [Rituxan]), and TNF-α (e.g., infliximab [Remicade])
Vaccine products	Hepatitis B surface antigen (e.g., Recomblvax HB), Borrelia burgdorferi outer sourface protein A (LYMErix), + human papillomavirus major capsid proteins (Gardasil)
Enzymes	Glucocerebrosidase (Cerezyme), DNase (Pulmozyme), thrombolytics (e.g., alteplase [Activase]), urate oxidase (rasburicase, Elitek)

TABLE 17.1   *(Continued)*

CLASS	EXAMPLES
Novel synthetic proteins	Fusion protein of interleukin-2 and diphtheria toxin (denileukin diftitox [Ontak]), soluble TNF receptor linked to IgG Fc (etanercept [Enbrel])
Novel conjugates	Pegylated proteins: interferon (peginterferon alfa-2a [Pegasys]), granulocyte colony-stimulating factor (pegfilgrastim [Neulasta]), human growth hormone (pegvisomant [Sornavert]) Covalently attached metal chelators: ibrutumomab tiuxetan (Zevalin) Covalently attached radioactive iodine: Iodine-131 tositumomab (Bexxar) Covalently attached chemotherapeutics: gemtuzumab ozogarnicin (Mylotarg)

(Courtesy: Paul Saenger, Current Status of Biosimilar Growth Hormone, International Journal of Pediatric Endocrinology, Volume 2009, Article ID 370329, 1–6)

TABLE 17.2   Definitions of biological and chemical pharmaceuticals

Generic drug	Chemical and therapeutic equivalent of a low-molecular-weight drug whose patent has expired
Biopharmaceutical	'A medicinal product developed by means of one or more of the following biotechnology practices: rDNA, controlled gene expression, antibody methods'
Biosimilar	'A biological medicinal product referring to an existing one and submitted to regulatory authorities for marketing authorization by an independent application after the time of the protection of the data has expired for the original product'

17.2   THE DRIVING FORCE BEHIND BIOSIMILARS

With the ever-increasing cost of pharmaceuticals, both for small molecules and biosimilars, there is an impetus to reduce the fiscal cost of these interventions to increase patient access and limit the rapidly expanding health-care budget. The arrival of generic medicines and attempts by regulatory authorities to cap costs by imposing significant reductions in reimbursement or price is a worldwide phenomenon. A number of biopharmaceutical patents are due to expire in the next few years (Refer Table 17.3), or have already expired such as human insulin, human growth hormone and interferon alfa and beta. The subsequent production of follow-on products, or 'biosimilars' has aroused interest within the pharmaceutical industry as biosimilar manufacturers strive to obtain part of an already large and rapidly-growing market. Demand for biologics is also growing rapidly. According to a 2009 Federal Trade Commission report, in 2007 American consumers spent ~$40 billion on biologics out of $287 billion spent for prescription drugs overall [5].

TABLE 17.3   Biopharmaceuticals that go Off-Patent in the Near Future

PRODUCT	GENERIC NAME	COMPANY	THERAPEUTIC SUBCATEGORY	PATENT EXPIRY
Rituxan	Rituximab	Roche	Antineoplastic, Mabs	31-12-2014
Gonal-F/Go-nalef	Follitropin alfa	Merck	KGaA Fertility agents	16-06-2015
Helixate	Octocog alfa	CSL	Antifibrinolytics	31-12-2015
Neulasta	Pegfilgrastim	Amgen	Immunostimulants	20-10-2015
Lantus insulin	glargine	Sanofi-Aventis	Antidiabetics	12-02-2015
Actemra	Tocilizumab	Roche	Antirheumatics	07-06-2015
Kogenate	Octocog alfa	Bayer	Antifibrinolytics	31-12-2014
Norditropin SimpleXx	Somatropin	Novo Nor-disk	Growth hormones	15-12-2015
Nimotuzumab	Nimotuzumab YM	BioSciences	Antineoplastic Mabs	17-11-2015
Prevnar	Pneumococcal vaccine	Pfizer	Vaccines	01-01-2015
Rituxan	Rituximab	Roche	Antineoplastic	31-12-2014

Courtesy: Sagar J. Kanase, Yogesh N. Gavhane, Ashok Khandekar, Atul S. Gurav and A.V. Yadav, Biosimilar: an overview, Kanase et al., IJPSR, 2013; Vol. 4(6): 2132–2144)

Manufacturers, policymakers and regulatory authorities must ensure that the economic benefits that biosimilars promise are not endangered by unique safety risks that biosimilars can pose.

17.3   DIFFERENCES BETWEEN GENERICS AND BIOSIMILARS

Biosimilars are fundamentally different from generic chemical drugs [6]. Generic drugs represent chemical and therapeutic equivalence to the original drug whose patents have expired. Most chemical drugs are low-molecular-weight compounds that are made from standard chemicals and reagents, involving organic chemistry [3]. In contrast, biopharmaceuticals are high molecular weight compounds with complex three-dimensional structures and the production process is much more complicated. This contrasting character differentiates chemical drugs and biopharmaceuticals is depicted in terms of their molecular weight (refer Table 17.4).

TABLE 17.4 Molecular weights of chemical drugs compared with biopharmaceuticals [6]:

	Drug Name	**Molecular Weight (Da)**
Chemical drugs	Paracetamol	151
	Rofecoxib	315
	Simvastatin	419
Biopharmaceuticals	Filgrastim	18 800
	Epoetin alfa	30 400
	Rituximab	145 000
	Factor VIII	264 000

It may be possible to mimic the initial amino acid sequence (primary structure), but further analysis of the biopharmaceutical's structure reveals both local folding (secondary structure) and subsequent global folding (tertiary structure), utilizing various hydrogen or disulfide bonds. Some biopharmaceuticals also exhibit quaternary structure, which is the stable association of two or more polypeptide chains into a multi-subunit structure and this may further alter activity, duration of action and other properties [7]. The various cell lines often involving heterogeneous mammalian cell lines that are used to produce the proteins may have an impact on the overall structure of the protein, and may affect post-translational modifications such as the extent of glycosylation. Alteration of the degree of glycosylation of molecules may impact greatly on receptor binding, both in duration and efficacy, in addition to altering metabolic profiles. Further modification may occur outside of cell culture, with the addition of polyethylene glycol bridges (pegylation), to join proteins and further alter receptor binding and subsequent metabolic removal [8]. In fact there is a higher barrier to entry for the biosimilar market than for small-molecule generics which includes higher costs, greater risks, greater time and expertise in relation to the clinical development of these products. Furthermore, the marketing and launch of biosimilars requires a different strategy than small-molecule generics (the strategy is described in Figure 17.1) [9].

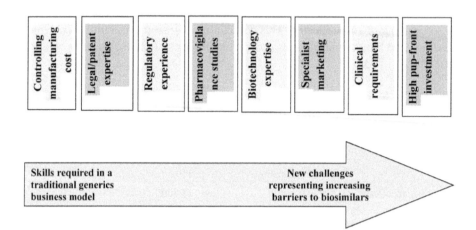

FIGURE 17.1   Strategy require to develop biosimilars [9]:

17.4   ISSUES WITH MANUFACTURING PROCESS FOR BIOSIMILARS

As already mentioned, the manufacturing process of biopharmaceuticals is complex and risky. Compared to the manufacture of small molecular entities, the manufacture of biopharmaceuticals requires a greater number of batch records (>250 versus <10); more product quality tests (>2,000 versus <100); more critical process steps (>5,000 versus <100) and more process data entries (>60 000 versus<4,000) [10]. The initial DNA sequence of the desired protein product must be determined and subsequently inserted into a suitable vector and then into the appropriate cell line. A bank of cells is derived and from that a master cell bank is defined. Cell culture of the master bank results in replication of cells and increased protein production. During extraction of particular protein, the supernatant (containing the protein) subsequently needs to be purified, which involves significant protein wastage to ensure adequate purity. No two master cell banks are the same and this account for the major dissimilarities between the innovator's product and the biosimilars [3].

The total production process is prone to variability and it is probably justifiable to say that it will be impossible to produce an identical molecule. Slight changes in production conditions can result in subtle changes in the final product and hence it is argued that they can never be bio-identical. This may be a result of inter- and intrabatch cell variability, different preservatives, and unsecured cold chain handing. Moreover there are major concerns over ensuring sterility and stability of the product delivered to market.

However, biotechnology has advanced to such an extent that in some cases it may be relatively easy for potential biosimilar manufacturers to create accurate copies of an innovator's biologic, by using microbial cell production rather than mammalian cell lines [8].

17.4.1   THE IMPACT OF DIFFERENCES

The impact of even small structural differences on clinically relevant properties of proteins may be significant. Each of these processes differences can impact on the interaction between the biopharmaceutical and the cellular receptor. In turn, these differences may lead to differences in efficacy and more importantly, their ability to trigger and then damage patient immune responses [6]. For glycosylated proteins, differences in the glycopattern may significantly influence receptor binding, protein–protein interaction and pharmacokinetics of protein substances [11]. In case of immunoglobulins, small differences in core fucosylation can lead to big changes in Fcc receptor binding and consequently result in changes of immune effector functions such as antibody-dependent cellular cytotoxicity which is believed to be a major mechanism of action contributing to the potency of many monoclonal antibodies, particularly in oncological indications [12].

17.5   SAFETY ISSUES-IMMUNOGENICITY

A key safety parameter in biopharmaceuticals is immunogenicity, i.e. the ability of a substance to trigger an immune response in the patient. Nearly all biopharmaceuticals induce antibodies, due to either the presence of foreign sequences or epitopes or the breaking of immune tolerance to self-antigens [13]. The latter mechanism which is not completely understood apparently does not only depend on patient characteristics, route of administration and disease-related factors but also on the quality of the protein product which includes the presence or absence of glycosylation, impurities as well as product handling issues have been associated with the induction of antibodies [14]. Therefore, products from different sources cannot be assumed to be equivalent concerning their immunogenic potential. There are various potential consequences of immunogenicity such as loss or enhancement of efficacy, neutralization of a native protein and general immune effects (allergy, anaphylaxis, serum sickness) [15]. There can be dramatic effects if a natural human protein with an essential activity is neutralized. Such cases had been described some years ago for an erythropoietin product where a postapproval formulation change led to an increased number of cases of pure red cell aplasia (PRCA) this complication was manifested by severe epoetin-resistant anemia which required blood transfusions, immunosuppressive treatment and eventually kidney transplantation [15], as well as for megakaryocyte-derived growth factor where severe thrombocytopenia was induced in volunteers and cancer patients, and led to the termination of product development [16]. In 2006, the German pharmaceutical company TeGenero conducted a Phase I clinical trial to test the safety of an experimental immuno-modulatory antibody developed to treat B-cell chronic lymphocyte leukemia and rheumatoid arthritis, known as TGN1412. In all participants, infusion with TGN1412 triggered the sudden release of proinflammatory cytokines, causing a condition known as system inflammatory response syndrome (SIRS) [17]. In a study of 174 biologics approved for use in the United States and Europe, postmarketing concerns were raised for nearly a quarter of those drugs. Safety related regulatory actions were issued for 41 biologics, including "black box" warnings on 19 biosimilars [18]. A list of approved and rejected biosimilars from 2006 is depicted in Table 17.5. Unfortunately, immunogenicity in humans is not predictable based on in-vitro or animal tests so that always data generated by clinical testing are required for assessment of immunogenicity and appropriate pharmacovigilance is mandatory.

TABLE 17.5   List of Approved and Rejected Biosimilar from 2006

BIOSIMILAR	REFERENCE	APPROVAL/REJECTION YEAR
Omnitrope	Somatropin	2006[a]
Valtropin	Somatropin	2006[a]
Binocrit	Epoetin alpha	2007[a]
Epoetin alpha Hexal	Epoetin alpha	2007[a]
Abseamed	Epoetin alpha	2007[a]
Silapo	Epoetin zeta	2007[a]
retacrit	Epoetin zeta	2007[a]
Filgrastim	Filgrastim	2008[a]
ratiopharm ratiograstim	Filgrastim	2008[a]
Biograstim	Filgrastim	2008[a]
Tevagrastim	Filgrastim	2008[a]
Filgrastim hexal	Filgrastim	2009[a]
Zarzio	Filgrastim	2009[a]
Nivestim	Filgrastim	2010[a]
Alpheon	Roferon-A	2006[b]
Humulin	Human insulin	2007[b]

[a]Approved, [b]Rejected

(Courtesy: Bhupinder Singh Sekhon, Vikrant Saluja, Biosimilars: an overview, Biosimilars 2011:1, 1–11)

17.6 LEGAL AND REGULATORY ISSUES WITH BIOSIMILARS

The "process equals product" paradigm emphasizes the importance of process control, process validation and product testing to overcome the differences in the product attributes caused due to differential manufacturing processes as they cannot be solely assessed by analytical characterization as recognized by the regulatory authorities [19]. Hence, specific regulatory pathways for licensing biosimilar medicinal products have been adopted in some parts of the world, where "generic pathway" (used for conventional small molecules) is not applicable for biopharmaceuticals.

Preliminary guidelines from the European Agency for the Evaluation of Medicinal Products (EMEA) states that the complexity of the product, the types of changes in the manufacturing process and differences in quality, safety and efficacy must be taken into account when evaluating biosimilars. For most products, results of clinical trials demonstrating safety and efficacy are likely to be required. Moreover, because of the unpredictability of the onset and incidence of immunogenicity, extended postmarketing surveillance is also important.

According to EMEA, all applications should be made entirely in accordance with the Common Technical Document (CTD) presentation. The CTD is organized into five modules that require being adapted for biosimilars as explained in following Table 17.6. The information to be supplied shall not be limited to Modules 1, 2 and 3 (pharmaceutical, chemical and biological data) supplemented with bioequivalence and bioavailability data. As commented before, the type and amount of additional data needs to be determined on a case-by-case basis studying the relevant scientific guidelines [20].

TABLE 17.6 Format of the dossier—modules of the Common Technical Document [20]:

Module 1	Administrative information	Normal requirements
Module 2	Quality, non-clinical and clinical summaries and overview	Normal requirements
Module 3	Quality (chemical, pharmaceutical and biological information)	Full + *comparability exercise*
Module 4	Non-clinical study reports	Reduced + *comparability exercise*
Module 5	Clinical study reports	Reduced + *comparability exercise*

This Directive indicates that comparability studies between the biosimilar and the reference medicine have to be performed but it does not address the requirements for such tests. These studies should be conducted at different levels.

—Physicochemical comparability.
—Biological comparability.
—Preclinical comparability.
—Clinical comparability.

The selected reference product will need to be the same throughout the comparability programme. Such comparability studies involve a thorough process starting by the comparison in terms of product quality and manufacturing process consistency as the safety and efficacy profile of the product is closely linked to its manufacturing method. Currently, due to the state of the art in science it is almost impossible to prove that two biologic medicines have the same qualitative and quantitative composition. In order to prove that there are no relevant differences between both medicines, in most, if not all cases, comparison to the reference product has to be performed at nonclinical level. In all cases there should be PK/PD (Pharmacokinetic/Pharmacodynamic) comparison of a biosimilar and reference product and in some cases clinical therapeutic equivalence trials are requested to show similar efficacy and safety at least in one clinical situation.

The applicant has to justify with regard to safety and efficacy of the drug and may have consequences as to the amount of nonclinical and clinical data to be provided. For those product classes for which guideline annexes are available at present, relatively simple, easy-to-measure clinical endpoints or accepted surrogate endpoints are available which facilitate comparative trials (refer Figure 17.2 for phases). In future, with products requiring more complex clinical endpoints (e.g., monoclonal antibody products), the design of the comparative equivalence trials may become much more challenging. Furthermore, as the differences are not fully apparent at the time of approval, the guidelines request that for biosimilars (as for all biological medicinal products) pharmacovigilance monitoring has to be performed [21]. For this purpose, the specific product given to the patient should be clearly identified.

Using this regulatory framework, a number of biosimilar products already have been licensed in the European Union and are being marketed in several, but not yet in all, European countries. Details on the various guidelines on biosimilars and the product data which led CHMP to recommend their approval can be found in the European Public Assessment Reports accessible via the EMEA webpage [22]. At launch, these products were offered about 15–35% price discount vs. the list prices of the innovator products (depending on the product, country, and package size). On the other hand, one interferon alfa product has been rejected by CHMP[23] and the applications for three insulin products have been withdrawn [24] demonstrating that the European regulations, in order to ensure a high standard of quality, safety and efficacy represent significant hurdles as appropriate to prevent the market entry of substandard products.

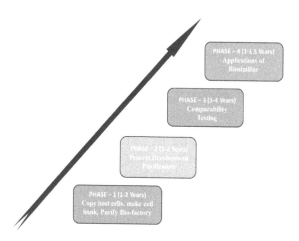

FIGURE 17.2   Regulatory approval duration

In 2006, Australia's Therapeutic Goods Administration (TGA) adopted the European Union's guidelines on the approval of biosimilars. Since then, regulatory authorities for other countries including Argentina, India, Japan, Mexico and Turkey have also issued draft or final guidelines on the issue[25].

In 2009, the World Health Organization's Expert Committee on Biological Standardization issued its Guidelines on Evaluation of Similar Biotherapeutic Products, according to which the guidelines provide "globally acceptable principles" for the approval of biosimilar products and can be adopted or used by regulatory authorities around the world in establishing regulatory frameworks for the approval of these products[26].

In USA, Biologics Price Competition and Innovation Act of 2009 (BPCI Act) establishes an abbreviated approval pathway for biological products that are demonstrated to be 'highly similar' (biosimilar) to, or "interchangeable" with, a FDA-licensed biological product. Under the BPCIA, a biosimilar product is "highly similar to the reference product not withstanding minor differences in clinically inactive components" with "no clinically meaningful differences" between the two products with respect to the "safety, purity, and potency of the product." The level of data required to demonstrate "highly similar," "minor differences" and "meaningful" may make all the difference and has yet to be determined [25]. USFDA had planned to establish authoritarian guidelines for biosimilars in 2011.

17.7   OPEN ISSUES WITH BIOSIMILARS

Although regulatory frameworks for biosimilars have been adopted or are up coming, in many parts of the world, there are some open issues left which are presently under intense discussion.

17.7.1   REFERENCE PRODUCT

According to the EMEA guidelines (depicted in Figure 17.3), the reference product has to be authorized in the EU, based on a full dossier, and the same reference product has to be used throughout the comparative studies. Data generated from studies with medicinal products authorized outside the community may only provide supportive information [27]. However, innovator products authorized in different countries may differ concerning (e.g., production site, formulation, and strength), so if the same demand would be made for all countries, a biosimilars manufacturer may be faced with the need to do comparative studies separately for each country. Therefore, the option of national regulatory authorities to accept a reference product not licensed within their jurisdiction is under discussion but would call for information sharing between the regulatory authorities and/or additional data to be provided by the biosimilars manufacturers.

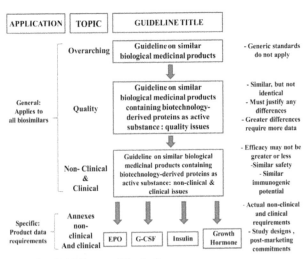

FIGURE 17.3   EMEA Guidelines on Biosimilars

17.7.2   LABELLING

Labelling for biosimilars is not an easy process when compared to generic small-molecule drugs because they are not identical, but only similar to their reference products and are licensed on the basis of their own development, including clinical data. Therefore, the labeling should differentiate clinical safety and efficacy data which have been obtained with the biosimilar product itself from reference product, particularly in extrapolated indications where no studies have been done with the biosimilars at all [28]. Furthermore unique safety data should be included and substitution advice should be provided.

17.7.3   PHARMACOVIGILANCE

In case of biosimilars, an appropriate system of pharmacovigilance is needed to assure responsibility for their products on the market and to ensure that appropriate action is taken if necessary [29] because the preauthorization safety database will be relatively small due to the abridged clinical development program. Pharmacovigilance is of special importance in case of rare serious adverse events which might not be evident at approval due to the limited data package available at this time. Pharmacovigilance systems based on spontaneous reporting will be limited by under-reporting as well as by data quality, which is often insufficient to allow a meaningful assessment [30]. Therefore, a more proactive approach may be required.

17.7.4   NAMING

In order to support postapproval monitoring, the specific medicinal product given to the patient must be clearly identified [19]. International nonproprietary names (INNs) are assigned to drug substances by the WHO INN Programme. The INN is the 'technical' name for medicinal products. The generic versions of chemical medicines are assigned the same name, as they are identical copies of the reference product [15]. WHO does not intend to introduce a specific process for naming biosimilars [31], and the INN as a cataloging system for drug substances can neither be relied upon as an appropriate means to ensure identification and traceability of biological, including biosimilar products nor as the sole indicator of product interchangeability. Therefore, it will be necessary that biosimilar products are marketed using brand names.

17.7.5   INTERCHANGEABILITY AND SUBSTITUTION

Small molecular generics can be interchangeable while biosimilars are not: here interchangeability should be demonstrated by scientific data proving that two products can be safely substituted for one another and do not create adverse health outcomes (e.g., generating a pathologic immune response after repeated switching). In the absence of such data patients and physicians should be aware that protein products with similar molecular composition may indeed not be interchangeable [32]. EMEA recommends that the decision to treat a patient with a reference or a biosimilar medicine should be taken following the opinion of a qualified healthcare professional [33].

17.8   CHECKPOSTS IN BIOSIMILAR MANUFACTURING

Development of a biosimilar molecule involves comparing the properties of biosimilar drug molecules with that of a reference product. The goal of development is to match all the known characteristics of the reference drug formulation. The prime motto will be to obtain all the genuine information about the reference formulation so as to ensure proper design, manufacture, and comparability of the biosimilar. Essential information of manufacturing aspects are kept hidden (e.g., process optimization).

Biosimilars are produced by biosynthetic processes using a biofactory (e.g., bacteria suffer from clonal mutations which affect batch process). One must be able to ensure that the product is of high consistency as desired. In comparison, molecular drugs are easily manufactured by synthesis. Due to these production differences, the product of the chemical synthesis will be pure and homogenous while the biopharmaceutical product will be well defined yet heterogeneous. This shows any biosimilar of active ingredients can be standardized easily compared to its protein counterpart. A pin-head change in the manufacturing conditions of a biopharmaceutical will affect quality, safety, and efficacy of the final product, which are the for-most criteria by ICH Guidelines. For biopharmaceuticals, most regulatory authorities require that not only the product meet defined criteria, is reviewed and approved, but also the manufacturing process and facilities meet requirements for consistent production. Thus, biosimilar product manufacturers must ensure their manufacturing process is consistently producing a biosimilar candidate that is comparable to the reference product [34].

17.9   BIOSIMILARS FOR PARTICULAR THERAPY

17.9.1   BIOSIMILARS FOR HEMATOLOGY

Filgrastim, a 175 amino acid recombinant-protein biosimilar reported to produce immediate transient leucopenia followed by increase in neutrophils population. Its activity is useful in treatment of Neutropenia. It was approved in Europe & USA and marketed under tradename 'Neupogen'. Another Biosimilar of Filgrastim is marketed by Hospira Ltd. under tradename "Nivestim".

Anaemia is another disease that is common among women (16–42 year). The treatment involves erythrocyte transfusion. For the treatment, one may use Erythropoitein as an approach. The available brands of biosimilars are "Retacrit" which is equivalent to "Epoetin alfa" [35].

17.9.2   BIOSIMILAR FOR GROWTH HORMONE

Following guidelines published first by EMEA in 2006 both Valtropin and Omnitrope have chosen the most sensitive model (prepubertal children with GH deficiency) with stable thyroid hormone and/or glucocorticoid replacement therapies if indicated, to show comparative efficacy and safety to the reference biological medicinal product.

Somatropin: The formulations include Sandoz powder for solution for injection (Covance). Somatropin Sandoz powder (Sandoz): formulation to be marketed as Omnitrope. Other formulation in liquid dosage form for adolescent is Somatropin Sandoz liquid (Sandoz) [36].

17.9.3   BIOSIMILAR FOR INSULIN

The major challenge is the method of production. For example, different laboratories use slightly different strains of bacteria and yeast that often leads to the production of nonidentical insulins, which will definitely have different clinical effects. A very small difference will have a significant difference in safety and efficacy along with receptor binding, duration of action, bioavailability and toxicity. Manufacturing related impurities are different for different brands during production which requires precise downstreaming. An important issue surrounding the insulin development is the drug delivery system [37].

17.9.4   BIOSIMILAR FOR RHEUMATOID ARTHRITIS

The biosimilars used in the treatment of Rheumatoid Arthritis involves use of monoclonal antibodies. Many reference products (eg; Rituximab) are formulated by many renowned manufacturers like Pfizer, Teva, Dr. Reddy's Ltd, Probiomed, etc. Different aspects are encountered while the development of MAbs for RA [38]. These include:

17.9.4.1   MANUFACTURING AND FUNCTIONAL IMPLICATIONS

Manufacture of large, complex proteins utilizes living cell cultures that are processed in highly controlled manner. A minute change in protein conformation may result in abolition of activity, insolubility and high immunogenicity, thus, amino acid sequences and higher-order structures must be same from batch to batch. Biological agents include protein formulations, such as hormones and somatostatins, with a molecular weight of ~5,000–25,000 Da. These exhibit a well-characterized structural feature that can be replicated using recombinant techniques. By contrast, mAbs are complex molecules, ~150 000 Da, which must be folded correctly to maintain conformational integrity. A posttranslational modification (e.g., glycosylation, methylation, oxidation, deamidation) influences tertiary and quaternary structures. Conformational integrity determines affinity, selectivity, functional activity and immunogenicity of mAbs, While these issues have generally led to concerns regarding inferiority of biosimilars compared with reference products, it must be borne in mind that such alterations could potentially lead to superior efficacy and safety. Thus biosimilar manufacturers must ensure sufficient analyses are performed to demonstrate a high degree of similarity between reference agents and biosimilars.

17.9.4.1   FC EFFECTOR FUNCTION

Activity of mAbs depends on Fc receptor (FcγR) function. Mutations produced by any factor will impair Fc interaction; therefore the complement activation, and reduced efficacy of mAbs. For example rituximab, display different levels of B-cell depletion, potentially due to altered fucosylation patterns. Due to constraints in conformational changes, etanercept exhibits reduced complement binding compared with infliximab and adalimumab. Efficacy of mAbs also suffers from paient to patient: RA and psoriatic arthritis, FcγR polymorphisms result in different responses to TNFi. Biosimilars must, therefore, demonstrate highly similar efficacy and safety to the reference product.

17.10   CONCLUSION

The market for biotechnology-derived medicinal products is evolving rapidly with the imminent entry of biosimilars. Product quality, safety and efficacy of biopharmaceuticals are highly reliant on the process of production, purification and formulation and subtle differences are often observed between the innovator's product and biosimilars. Therefore, it is important to show that biosimilar drugs are comparable in structure and function to that of the innovator and any differences have to be supported with data showing no influence on these parameters. The only way to ascertain the safety and efficacy of a biosimilar will be to conduct preclinical tests and clinical trials and implement tailored pharmacovigilance plans. Awareness of the differences between original biotechnological medicines and biosimilars is essential for the safety of the patients.

KEYWORDS

- **Biopharmaceuticals**
- **Biosimilars**
- **EMEAGenerics**
- **Immunogenicity**

REFERENCES

1. Schellekens, H.; Biosimilar therapeutics-what do we need to consider?. *Nephrol. Dialysis. Transplan. Plus.* **2009**, *2*(Suppl 1), i27–i36.
2. Schellekens, H.; Follow-on biologics: challenges of the 'next generation'. *Nephrol. Dial . Transpl.* **2005**, *20*(Suppl 4), iv31–iv36.
3. Roger, S. D.; Biosimilars: How similar or dissimilar are they? *Nephrology.* **2006**, *11*, 341–346.
4. Crommelin, D. J.; Bermejo, T.; and Bissig, M.; Pharmaceutical evaluation of biosimilars: important differences from generic low-molecular weight pharm. *Eur. J. Hosp. Pharm. Sci.* **2005**, *1*, 11–17.
5. Federal Trade Commission (FTC). Emerging Health Care Issues: Follow-On Biologic Drug Competition: A Federal Trade Commission Report. Washington, DC; **2009**.
6. Roger, S. D.; and Ashraf, M.; Biosimilars: opportunity or cause for concern?. *J. Pharm. Pharmaceut. Sci.* **2007**, *10*(3), 405–410.
7. Horton, H. R.; Moran, L. A.; Ochs, R. S.; Rawn, D.J.; and Scrimgeour, K. G.; Principles of Biochemistry. 3rd ed. New Jersey: Prentice Hall/Pearson Education; **2002**.
8. Wildt, S., and Gerngross, T. The humanization of N-glycosylation pathways in yeast. *Microbiology*, **2005**, *3*, 119–128.
9. Mark, J. B.; Laura, M. Harris; Romita, R. D.; and Chertkow; J.; Biosimilars: initial excitement gives way to reality. From the analyst's couch 2006 July, 5, 535–536.
10. Schellekens, H.; Manufacture and quality control of biopharmaceuticals. Revista Espa˜nola de Econom´ıa de la Salud, **2008**, *6*, 340–344.
11. Lis, H.; and Sharon, N.; Protein glycosylation-structural and functional aspects. *Eur. J. Biochem.* **1993**, *218*, 1–27.

12. Arnold, J. A.; Wormald, M. R.; Sim, R. B.; Rudd, P. M.; and Dwek, R A.; The impact of glycosylation on the biological function and structure of human immunoglobulins. *Annu. Rev. Immunol.* **2007**, *252*, 1–50.

13. Kresse, G. B.; Biosimilars-Science, status, and strategic perspective. *Eur. J. Pharm. Biopharm.* **2009**, *72*, 479–486.

14. Mukovouzov, I.; Sabljic, T.; Hortelano, G.; and Ofosu, F. A.; Factors that contribute to the immunogenicity of therapeutic proteins. *Thromb. Haemost.* **2008**, *99*, 874–882.

15. Nowicki, M.; Basic facts about biosimilars. *Kidney. Blood. Press. Res.* **2007**, *30*, 267–272.

16. Basser, R. L.; O'Flaherty E.; Green, M.; Edmonds, M.; Nichol, J.; Menchaca, D. M.; Cohen, B,; and Begley, C. G.; Development of pancytopenia with neutralizing antibodies to thrombopoietin after multicycle chemotherapy supported by megakaryocyte growth and development factor. *Blood.* **2002**, *99*, 2599–2602.

17. Suntharalingam G.; Cytokine Storm in a Phase 1 Trial of the Anti-CD28 Monoclonal Antibody TGN1412. *New Engl. J. Med.* **2006**, *355*(10), 1018–1028.

18. Giezen, T. J.; Safety-Related Regulatory Actions for Biologicals Approved in the United States and the European Union. *JAMA.* **2008**, *300*(16), 1887–1896.

19. Andrzej, W.; and Ashraf, M.; European regulatory guidelines for biosimilars. *Nephrol. Dial. Transplant.* **2006**, *21*(Supp 5), v17–v20.

20. Leyre, Z.; and Begoña, C.; Biosimilars approval process. Regulatory Toxicology and Pharmacology. **2010**, *56*, 374–377.

21. EMEA; Guideline on similar medicinal products containing biotechnology derived proteins as active substance: non-clinical and clinical issues,EMEA/CHMP/BMWP/ 42832/2005, 2006. Available from: <http://www.emea.europa.eu/pdfs/human/biosimilar/4283205en.pdf>.

22. EMEA; EPARs for authorised medicinal products for human use. Available from: <http://www.emea.europa.eu/htms/human/epar/eparintro.htm>.

23. EMEA; Refusal assessment report for Alpheon, EMEA/H/C/000585, 2006. Available from: <http://www.emea.europa.eu/humandocs/PDFs/EPAR/alpheon/H-585-RAR-en.pdf>.

24. EMEA; Withdrawal assessment report for Insulin Human Rapid Marvel,EMEA/H/C/845, **2008**. Available from: <http://www.emea.europa.eu/humandocs/PDFs/EPAR/ insulin humanrapidmarvel/31777807en.pdf>.

25. Madeleine, M.; and Jennifer Stonecipher, H.; Learning from our neighbors about regulation of biosimilar drugs. The national law journal, food and drug practice a special report 2011 Mar 14.

26. Barbara Mounho.; Global Regulatory Standards for the Approval of Biosimilars, 65 Food & Drug L. J. **2010**, 819.

27. EMEA; Annex to guideline on similar biological medicinal products containing biotechnology-derived proteins as active substance: non-clinical and clinical issues/Guidance on similar medicinal products containing recombinant erythropoietins, EMEA/CHMP/BMWP/94526/2005, **2006**. Available from:<http://www.emea. europa.eu/ pdfs/human/biosimilar/9452605en.pdf>.

28. European Commission, Notice to applicants: a guideline on summary of product characteristics, 2005. Available from: <http://ec.europa.eu/enterprise/pharmaceuticals/ eudralex/vol-2/c/spcguidrev1-oct2005.pdf>.

29. EMEA; CPMP Position Paper on Compliance with Pharmacovigilance Regulatory Obligations, CPMP/PhVWP/1618/01, 2001. Available from <http://www.emea. europa.eu/ pdfs/ human/phvwp/161801en.pdf>.

30. EMEA; Executive Summary Report on EMEA Meeting with interested parties and research centers on ENCEPP, EMEA/601107/2007, 2007. Available from: <http://www.emea.europa. eu/pdfs/human/phv/60110707en.pdf>.

31. WHO Informal Consultation on International Nonproprietary Names (INN) Policy for Biosimilar Products, Geneva, 4–5 September **2006**. Available from: <http://www.who.int/medicines/services/inn/BiosimilarsINN_Report.pdf>.

32. EMEA; Guideline on similar medicinal products containing biotechnology derived proteins as active substance: quality issues, EMEA/CHMP/BWP/49348/2005, **2006**. Available from: <http://www.emea.europa.eu/pdfs/human/ biosimilar/4934805en.pdf>.

33. EMEA, Questions and Answers on biosimilar medicines (similar biological medicinal products), EMEA/74562/2006, 2007. Available from: <http: //www.emea.europa.eu/pdfs/human/ pcwp/7456206en.pdf>

34. Williams, B. R.; and Shen, D. W.; Current Considerations for Biosimilar Therapeutics. *Trend. Bio/Pharmac. Ind*, 12–13.

35. Waller, C. F.; Biosimilars and their use in hematology and oncology. *Community Oncol.* **2012**, *9*, 198–205.

36. Saenger P. Current Status of Biosimilar Growth Hormone. International Journal of Pediatric Endocrinology. Volume 2009, Article ID 370329, 3-4.

37. Gough, S.; Biosimilar insulins: opportunities and challenges. *Pract. Diab. 30*(4) 146–147.

38. Dörner, T.; Strand, V.; Castañeda-Hernández, G.; Ferraccioli, G.; Isaacs, J. D.; Kvien T.K.; Martin-Mola, E.; Mittendorf, T.; Smolen, J. S.; and Burmester, G. R.; The Role of Biosimilars in the Treatment of Rheumatic Diseases. *Ann. Rheum. Dis.* **2013**, *72*(3), 322–328.

CHAPTER 18

APPLICATION EFFICIENCY OF MICROBIAL PREPARATION DESIGNED TO INTENSIFY DISPOSAL OF LIPID COMPOUNDS IN WASTEWATERS

M. S. CHIRIKOVA, T. P. SHAKUN, and A. S. SAMSONOVA

Institute of Microbiology, National Academy of Sciences, Belarus 220141, Kuprevich str.2, Minsk, Belarus

E-mail: margarita.chirikova@mail.ru, Fax: +375(17) 267-47-66

CONTENTS

18.1 INTRODUCTION

Nowadays pollution of aguatic systems with effluents containing various contaminants, including waste lipids has turned into priority problem. Organic components discharged into water reservoirs create favorable conditions for activities of pathogenic bacteria, fungi, and protozoa. They undergo complex biochemical transformations causing thereby secondary contamination and direct adverse effect on local biota [1].

Recovery of municipal sewage and production effluents became an acute challenge. Deregulated urban and industrial runoff interferes with performance of biological decontamination stations based on activated sludge technology. Microbial components of sludge biocenosis are not able to cope with elevated concentrations of the pollutants [2].

The functioning of aeration ponds is complicated by presence in water of huge amounts of lipid wastes separated by mechanical method. This, in turn, arouses the problem of burying the collected by-lipids [3].

In recent years biopreparations composed of microbial degraders are actively applied for treatment of lipid-saturated wastewaters [4]. Microorganisms synthesize lipolytic enzymes effectively breaking down the organic pollutants. Fats and oils are converted into ecologically harmless products [5].

Over 80 dairy plants are operating in Belarus, with overall output of lipidic effluents reaching 10–12 mln ton per year. Lack of home-made biopreparations, intensifying recovery of such wastewaters leads in some cases to illicit dumping and massive environmental damage.

Aim of study was to evaluate application efficiency of microbial preparation promoting bioremediation of wastewaters polluted with lipid compounds.

18.2 OBJECTS AND METHODS OF INVESTIGATION

Microbial preparation Antoil developed at the Institute of Microbiology, National Academy of Sciences, Belarus was chosen as the object of studies. The product incorporates active microbial strains-lipid degraders: *Rhodococcus sp.* R1-3FN, *Rhodococcus ruber* 2B, *Bacillus subtilis* 6/2-APF1, *Pseudomonas putida* 10AP.

Lipase activity was assayed by Ota-Yamada method [6]. It evaluates alkalititrated fatty acids released by lipase action from olive oil substrate. The difference between titration results of test and control samples corresponds to the amount of 0.05 n NaOH solution spent for neutralization of fatty acids produced by lipase in the course of olive oil treatment.

Lipolytic activity of microbial cultures LC (u/g) is calculated by formula 8.1:

$$LC = \frac{(A \cdot T - 50)}{B},$$ (18.1)

where LC—lipolytic activity, u/g; A—difference between titration results in test and control samples, cm³; T—alkali titer; B—concentration of enzyme solution sample, g/cm³.

Degradation of lipid substances was studied in 500 ml Erlenmeyer flasks containing 100 ml of mineral medium E8 supplemented with 0.1 percent lipid substrates as nutrition sources (lard, milk fat, sunflower and olive oil). Aerobic fermentation was carried out on orbital shaker at 150 rpm and temperature 28°C. Each flask was seeded with 10 percent inoculum (v/v).

Lipid amount in the sample was measured gravimetrically [7].

Fats, oils, other lipids, crude oil fractions were recovered from wastewater by multiple petroleum ether extraction. The resulting extract was divided into 2 parts.

In one portion the solvent was evaporated and the residue was weighed to find the total content of lipids and nonvolatile petroleum products according to formula 18.2:

$$x1 = \frac{(m1 - m2) \cdot V2 \cdot 1000}{V1 \cdot V}$$ (18.2)

where $x1$—total concentration of substances extracted with petroleum ether, mg/l; $m1$—mass of weighing bottle with residue remaining after evaporation of extractive agent, mg; $m2$—net bottle weight, mg; $V1$—extract aliquot volume, ml; $V2$—volume of graduated flask with extract, ml; V—volume of analyzed sample, ml.

The other portion of the extract was passed through aluminum oxide and the content of petroleum products was assessed gravimetrically by formula 18.2. The difference in two values constituted actual concentration of fats and other lipids extracted with petroleum ether.

Large-scale mixed culture of bacterial strains *Bacillus subtilis* 6/2-APF1 and *Pseudomonas putida* 10AP was conducted in 300 liter fermentor LiFlus SP on Meynell nutrient medium at temperature 28 ± 2°C during 48 hrs. Mixed fermentation of bacterial strains *Rhodococcus sp.* R1-3FN and *Rhodococcus ruber* 2B was performed in fermentor LiFlus SP (300 l) on Meynell medium at temperature 28±2°C during 98 hrs.

Chemical oxygen demand (COD) was determined by express method [7].

18.3   RESULTS AND DISCUSSION

Screening of microorganisms for the ability to utilize fats and oils as nutrition sources embraced 40 cultures collected by researchers from laboratory of xenobiotic degradation and bioremediation of natural and industrial media. Collection entries were isolated from municipal sewage and activated sludge of decontamination stations treating wastewaters of organic synthesis and dairy/meat processing overloaded with fatty substances.

Degrading potential of tested microbial species was judged by the ability to grow on synthetic medium E8 containing 1percent concentration of lipids supplied from vegetable oils-sunflower, olive and animal fats—pork and milk.

Overwhelming majority of cultures capable to grow on media with oils and fats is represented by genus *Rhodococcus*. It accounted for 67.5 percent among 40 examined isolates. 20 percent share belongs to *Bacillus* genus, the rest 12.5 percent is occupied by *Pseudomonas* genus.

Ten cultures displayed most active growth, with 4 strains taking the dominant lead: *Rhodococcus sp.* R1-3FN, *Rhodococcus ruber*2B, *Bacillus subtilis* 6/2-APF1 and *Pseudomonas putida* 10AP.

Maximal lipolytic activity of microorganisms consuming lipids as a sole source of nutrition revealed in cultural liquid of selected superactive variants varied from 0.65 to 0.70 u/mg. Strains with inferior lipolytic activity showed the values in the range 0.3—0.6 u/mg (Table 18.1).

TABLE 18.1   Lipase activity of tested microorganisms

№	Strains	Specific Activity, u/mg Protein
1	*Bacillus subtilis* 6/2-APF1	$0.68 \pm 0{,}04$
2	*Pseudomonas putida* 10AP	$0.65 \pm 0{,}04$
3	*Pseudomonas fluorescens* 12B	$0.30 \pm 0{,}05$
4	*Rhodococcus sp.* R1-3FN	$0.72 \pm 0{,}04$
5	*Rhodococcus erythropolis* 70F	$0.51 \pm 0{,}06$
6	*Rhodococcus erythropolis* 23F	$0.50 \pm 0{,}05$
7	*Rhodococcus opacus* 100B	$0.47 \pm 0{,}03$
8	*Rhodococcus opacus* 29D	$0.52 \pm 0{,}04$
9	*Rhodococcus ruber* 1B	$0.60 \pm 0{,}04$
10	*Rhodococcus ruber* 2B	$0.70 \pm 0{,}03$

Strains *Rhodococcus sp.* R1-3FN, *Rhodococcus ruber* 2B, *Bacillus subtilis* 6/2-APF1 and *Pseudomonas putida* 10AP were rated as the most promising ingredients of microbial preparation elaborated to accelerate removal of lipid pollutants from wastewaters.

Application prospects of aforecited strains in biotechnology of formulating remediation preparation are grounded on the ability of pure microbial cultures to utilize fats and oils as nutrition sources.

Decomposing activity of four selected microbial strains was evaluated in 500 ml Erlenmeyer flasks where 0.1 percent concentrations of lard, milk fat, olive and sunflower oil served as nutrient substrates. Fermentation proceeded for 7 days on orbital shaker at agitation rate 150 rpm. Upon 7 days residual lipid substances were determined in the cultural liquid (Table 18.2).

TABLE 18.2   Oil and fat degradation efficiency of Antoil microbial constituents

Strain	Decomposition degree, %			
	Lard	Milk fat	Sunflower oil	Olive oil
Rhodococcus ruber 2B	84.3	88.2	86.2	84.2
Rhodococcus sp. R1-3FN	90.2	90.9	82.4	88.2
Pseudomonas putida 10AP	91.9	89.2	86.8	87.4
Bacillus subtilis 6/2-APF1	92.5	94.2	86.1	90.4

Investigation of the ability of tested microbial variants to digest oils and fats as nutrients demonstrated that all four strains catabolized both types of substrates. Strains *Rhodococcus sp.* R1-3FN, *Bacillus subtilis* 6/2-APF1 preferred lard and milk fat. Strain *Pseudomonas putida 10 AP* found lard most appetizing. Utilization efficiency of lard, milk fat, sunflower and olive oil by *Rhodococcus ruber 2B* constituted 84.3 percent, 88.2 percent, 86.2 percent and 84.2 percent, respectively.

Industrial trials of Antoil performance were completed at decontamination unit of Kopyl dairy plant and at biological sewage processing facilities of Kopyl municipal public utility network.

Antoil preparation was fed into four segments of aeration tank treating effluents of Kopyl dairy plant in amount 100 l per each tunnel and run-off was ceased afterwards. Biopreparation was incubated with wastewaters at continuous aeration during 8 hrs.

Initial contamination level of effluents denoted as COD in 4 sectors of aeration tank prior to Antoil supply equaled 840, 860, 830, and 850 mg O_2/l, respectively.

After Antoil treatment COD values sharply declined to 440, 430, 400, 410 mg O_2/l, respectively. Vital COD indices fell by 47.6 percent, 50 percent, 51.8 percent, and 51.7 percent, respectively, making the average percentage 50.2 percent.

Concentration of fatty substances in effluents of Kopyl dairy plant before Antoil application reached 750 mg/l. In post-Antoil samples concentration of lipid pollutants was drastically reduced by 90.4 percent to 72 mg/l. The obtained data are summarized in Table 18.3 .

TABLE 18.3 Results of Antoil efficiency trials at decontamination facilities of Kopyl dairy plant

Experimental variant	COD values, mg O_2/l				Amount of lipid substances, mg/l
	1-t segment of aeration tank	2-d segment of aeration tank	3-d segment of aeration tank	4-th segment of aeration tank	
Before Antoil application	840	860	830	850	750
After Antoil treatment	440	430	400	410	72
Reduction of parameters, %	47.6	50	51.8	51.7	90.4

Decrease of pollution level of dairy effluents illustrated by 50.2 percent COD decline and decay of lipid substances by 90.4 percent laid the basis for further series of Antoil performance tests at biological detoxification station of Kopyl public utility network.

Successful testing program demonstrated that introduction of defatting preparation into aeration tank resulted in 87.6 percent reduction of COD level in treated wastewaters. Initial amount of lipid pollutants in effluents (48.8 mg/l) fell drastically to trace concentrations, i.e. decontamination efficiency reached 99.9 percent.

18.4   CONCLUSIONS

High degrading activity of Antoil microbial constituents (84.3–92.5 %) and efficiency of wastewater recovery in terms of COD reduction (87.6 %) and lipid decomposition (99.9 %) registered during industrial trials at Kopyl municipal decontamination station evidence its attractive application prospects for intensification of lipid disposal in effluents. Technological introduction of this microbial preparation into bioremediation schemes will be of considerable social significance because it will promote degradation of lipid pollutants in the discharged wastewaters and hence will improve ecological situation in Belarus.

KEYWORDS

- **Chemical oxygen demand**
- **Disposal efficiency.**
- **Intensification of decontamination**
- **Lipid substances**
- **Lipolytic activity**
- **mMicrobial degraders**
- **Mmicrobial preparation**

REFERENCES

1. Ivchatov, A. L.; and Glyadenov, S. N.; Another aspect of biological decontamination of effluents. *Ecol. Ind. Russia.* **2003**, *4*, 37–40 (in Russian).
2. Samsonova, A. S.; Glushen, E. M.; Chirikova, M. S.; and Petrova, G. M.; Microorganisms intensifying disposal of lipid substances in wastewaters. Microbial biotechnologies: basic and applied aspects. Belaruskaya navuka Press: Minsk, **2012**. p. 250–259 (in Russian).
3 Poskryakova, N. V.; Development of biopreparation basis for lipid degradation. Ufa, **2007**, 24 p. (in Russian).
4. Murzakov, B. G.; Zaikina, A. I.; Zobnina, V. P.; Listov, E. L.; Zorina, L. V.; Rogacheva, R. A.; Biotechnological method for removal of fats and oils from wastewaters. Russian patent №2161595, **2001**. 23 p. (in Russian).
5. Reimann, J.; and Gotsche, A.; Reinigung fetthaltiger Abwasser der Frostfischindustrie nut thermophilen Mikroorganismen. *Chem-Ind-Techn*, **2002**. *5*, 634.
6. Gerhard et al. (Eds) Methods of General Bacteriology. Mir, Moscow, **2001**, 536 p. (in Russian).
7. Lurie, YU YU, Analytical chemistry of industrial effluents. Chemistry series, Moscow, **2004**. 448 p. (in Russian).

CHAPTER 19

VOLATILE STEROIDS AS POTENTIAL REGULATORS OF REPRODUCTION IN THE HOUSE MOUSE

M. A. KLYUCHNIKOVA and V. V. VOZNESSENSKAYA

A.N. Severtsov Institute of Ecology and Evolution 33, Leninski prospect, Moscow, Russia, 119071, E-mail: klyuchnikova@gmail.com, Fax: (495)9545534

CONTENTS

19.1   INTRODUCTION

The House Mouse (*Mus musculus*) is a pest species that cause considerable economic damage to field and fruit crops. Zoonoses (diseases transmitted from rodents to people) also pose a definite threat for human health and safety since mice are closely associated with human habitat areas. The prerequisite for the development of pest management tools is maximizing the safety of the environment, humans and other animals. Methods widely used today to control rodent populations in Russia are highly toxic to humans and other nontarget mammalian species (Voznessenskaya et al., 2004; Voznessenskaya, Malanina, 2014). Major pitfalls of current approaches are high toxicity to humans and other nontarget species, environmental pollution, development of avoidance behavior, and rodenticide resistance in rodents. Furthermore, individuals who survived after rodenticide treatment often exhibit reproductive outbreaks (Rylnikov et al., 1992). In turn, rapid increases in rodent population size are associated with the spread of zoonoses attributed to rodents. In general, there are two major strategies for the animal pest population management, which are strict control, such as killing or removing offending animals, and a more mild prevention of pest population problems. The last mentioned was proved to be more applicable for species with high population turnover, like mice and rats. Under natural conditions rodent population size is regulated by a number of external and internal (zoosocial) factors. Predator population density is the most influential external factor (Hentonnen et al, 1987; Voznessenskaya, 2014). Major internal regulating factor is rodent population density itself. Both factors are to a large extent dependent on natural chemosensory cues and that has already found some practical application. For example, predator urine is used to repel herbivorous species. Social behavior of the house mouse is largely guided by odors. The aim of current study is to determine the role of potential chemical signals of steroid nature in mouse social behavior and hormonal responses, which are closely related to internal factors controlling population density. This research lays down a fundamental basis for the development of novel humane and environmentally friendly tools for rodent pest management.

Steroid hormone metabolites are suggested to be involved in chemical communication in mice (Ingersoll, Launay, 1986; Nodari et al., 2008; Voznessenskaya et al., 2010). Endogenous steroid hormones reflect physiological, reproductive and social status in mammals. Thus, their metabolites are likely to serve signal pheromones, convey information about sex, social rank and reproductive ability of the individual, and elicit behavioral or/and hormonal responses. A volatile steroid of gonadal origin, androstenone (5α-androst-16-en-3-one) was chosen as a model odorant in our studies. Androstenone is well known as a sex boar pheromone; it facilitates the lordosis response when detected by receptive sows

(Reed et al., 1974). Androstenone and related steroids are also suggested to be involved in human chemical communication. A number of reports suggest that these steroids may act as a sexual attractants, have specific effects on emotions (Cowley et al., 1977; Kirk-Smith et al., 1978; Benton, 1982; Van Toler et al., 1983; Filsinger et al., 1984, 1985; Pause, 2004). However, the results of these studies remain controversial (Wysocki and Preti, 2009). In mice, Ingersoll and Launay (1986) found that androstenone induce intermale aggression, depending on stimulus presentation. It provoked aggressive behaviors if being mixed with or applied in close proximity to castrated male urine, but not if being dissolved in water. Mouse olfactory system is subdivided into the main and accessory olfactory systems. The accessory olfactory (vomeronasal) system is considered to be more narrowly tuned on the detection of pheromones and pheromone-like substances, although pheromones and odorants can be processed by both systems (Sam, et al., 2001; Levai, et al., 2006). Behavioral and immunohistochemical data from our previous research indicate that both the main olfactory and vomeronasal systems are capable of detecting androstenone in mice (Voznessenskaya et al., 2010). This evidence suggests that androstenone may play a role in mouse chemical communication. Present paper examines further the role of the volatile steroid androstenone in chemical communication of the house mice.

19.2 MATERIALS AND METHODS

19.2.1 TEST SUBJECTS

For the experiments we used 3-6 month old male CBA/Lac mice obtained from the Andreevka animal nursery (Moscow Region, Russia) as well as adult *Mus Musculus* males trapped in natural habitats in Moscow Region, Russia. The CBA inbred strain was chosen as the most sensitive to the odor of androstenone among 28 mouse strains tested (Voznessenskaya, Wysocki, 1994; Voznessenskaya et al., 1995). Previously, olfactory thresholds to androstenone in adult CBA mice were determined in the three different behavioral training paradigms; their sensitivity ranged from 2.8×10^{-5} to 9.2×10^{-5} % (w/v) androstenone solution in mineral oil (Voznessenskaya et al., 1999). Mice were kept under standard laboratory conditions with a 12 h-12 h light-dark cycle (lights on at 8.00 am) and temperature 20–22°C. Pelleted food and water were available *ad libitum*. CBA/Lac males were sexually experienced and were housed in individual cages for at least two weeks before the experiments. Prior to experiments, wild male mice were maintained singly under the above described laboratory conditions for three-four weeks to allow them to adjust.

19.2.2   ANDROSTENONE SOLUTIONS

0.1 % (w/v) stock solution was prepared by dissolving 5α-androst-16-en-3-one (Sigma, USA) in odorless mineral oil (MP Biomedicals, USA).

19.2.3   EXPERIMENTAL DESIGN

Experiment 1A. The experiment served to investigate the effects of androstenone exposures on plasma testosterone and corticosterone levels. These steroid hormones are associated with many aspects of mouse social behavior. CBA/Lac males (n = 15) were randomly assigned to two groups. One group of animals was exposed to 5 µL of 0.1 percent androstenone for 30 min, while another control group was exposed to 5 µL of mineral oil (vehicle). The duration of odor presentation (30 min) was chosen according to the timing of classical testosterone surge in response to the chemical signals presentation from an estrous female (Macrides et al., 1975; Maruniak et al., 1976; Osadchuk, Naumenko, 1981). We collected blood for hormone assay immediately after exposures.

Experiment 1B. In this experiment we simulated the presence in the environment of two potentially competing signals: a signal of receptive female and a male-derived steroid signal (androstenone). Design of the experiment was analogous to the described above, differing only in presented odor samples. CBA/Lac (n=16) and wild *Mus Musculus* males (n = 25) were simultaneously exposed to two separate odor samples: one containing 5 µL of 0.1 percemt androstenone or 5 µL of mineral oil (vehicle) and the other containing 50 µL of estrous female BALB/c urine.

Experiment 2. This experiment was designed to investigate possible effects of androstenone on mouse behavior. CBA/Lac males (n=27) were randomly assigned to two groups, one of which was exposed to 160 µL of 0.025% androstenone and the other was presented with control odor - 160 µL of mineral oil. In 5 minutes after completion of odor exposures we performed standard odor preference test (estrous female urine versus male urine). Normally male mice of this strain spend significantly more time investigating estrous female odors versus diestrous female or male odors. This test, after another 5 min break, was followed by the hole-board procedure.

19.2.4   URINE COLLECTION

Male and female urine from BALB/c strain and outbred laboratory mouse population was used for odor preference tests and odor exposures. Urine was collected by gently applying pressure to the mouse abdomen. Shortly after procedure, urines from 6 to 8 animals of the same group were pooled together, mixed, aliquoted, and stored at −22°C until use. In females, stages of estrous cycles were determined by vaginal smear cytology before urine collection. Three BALB/c males were castrated under anesthesia with pentobarbital (30 mg/kg, i.p.) a month prior urine collection.

19.2.5   ODOR EXPOSURES

To prevent air contamination odor exposures were performed either in the individually ventilated compartments of animal cabinets (ASP 130 Flufrance, France) in Experiment 1 or under the fume hood in the Experiment 2. In Experiment 1 androstenone solution or mineral oil or estrous female urine was applied on the gauze tissue secured out of the reach of a mouse in a plastic perforated container (d = 2.5 cm). Each odor sample also contained 10 μL of castrated BALB/c male urine (not mixed with androstenone). This was done in compliance with the observations of Ingersoll and Launay (1986), who found that androstenone promoted aggressive behavior in male mice only if presented in close proximity to or mixed with castrated male urine. During odor presentations containers were placed on the bedding of individual home cages. In Experiment 2 androstenone solution or mineral oil was applied on the home cage bedding.

19.2.6   ASSAY FOR PLASMA TESTOSTERONE AND CORTICOSTERONE

Mouse blood was collected within 1–2 min after the end of experimental procedures into Lithium Heparin coated micro tubes (Sarstedt, Germany). Blood samples were centrifuged for 15 min at 12,000 rpm. Separated plasma was immediately frozen and stored at −22°C until analysis. Concentrations of hormones in blood plasma samples were determined by ELISA technique using commercially available kits (testosterone EIA1559 and corticosterone EIA4164, DRG, USA). SpectraMax 340PC spectrophotometer equipped with SpectraMax software was employed to measure absorbance and to calculate hormone concentrations (Molecular Devices, USA).

19.2.7   BEHAVIORAL TESTING

We examined preferences of CBA/Lac males for estrus female odors in a standard test. During the experiment animals remained in their home cages (25cm x 14cm x 12cm) while food and water were removed. To prepare odor samples

20 µL of estrous female or male urine was applied to a small cotton ball dipped inside 1.5 ml plastic micro tube so that the animal could not reach it. Estrous female and male odor samples were presented simultaneously to CBA/Lac male for 10 min. Odor-containing micro tubes with their openings facing the animal were attached to the metal grid of the cage at the 10 cm distance from each other. We recorded number of approaches to odor samples and total time of investigation including sniffing, licking, gnawing and touching with head or front paws each micro tube opening.

The orienting-investigatory reactions of CBA/Lac males were studied in the hole-board test for 10 min (Voznessenskaya, Poletaeva, 1987). An animal was placed in the center of the arena and behavior recording was immediately started. We registered the number of nose pokes into the holes, rearing (standing on hind paws), the number and duration of grooming acts, number of defecations and urinations.

The data from behavioral testing were analyzed using criteria of parametric and nonparametric statistic implemented in the Statistica 7.0 (StatSoft, Inc) and Microsoft Excel 2010 software. In the following text, we present our data as group mean ± SEM.

19.3   RESULTS

Experiment 1A. The influence of androstenone exposures on plasma testosterone and corticosterone levels.

Plasma testosterone in males exposed to 0.1% androstenone was significantly lower (1.1 ± 0.3 ng/mL, n=8) than in the control group (11.7 ± 3.5 ng/mL, n = 7, $p < 0.05$, Student's t-test for the case of unequal variances). Group means for the concentration of corticosterone for the same blood samples was not affected (75.5 ± 18.7 ng/mL in the experimental group, 78.3 ± 13.9 ng/mL in the control group, $p > 0.05$, Student's t-test).

Experiment 1B. The influence of androstenone exposures on male plasma testosterone level in the presence of estrous female odor.

30 min androstenone exposures (0.1%, 5 µL) resulted in a significant decrease of plasma testosterone level in CBA/Lac males (2.64 ± 0.56 ng/ml (n = 8) vs. 15.3 ± 0.78 ng/mL (n=8), $p < 0.001$, Mann–Whitney test). Meanwhile we did not observe statistically significant changes in hormone level of wild-trapped *Mus Musculus* males (Student's t-test, $p = 0.13$). The concentration of testosterone in blood plasma of these males after androstenone exposure (0.1%, 5 µL) was 0.4 ± 0.1 ng/ml (n=13), compared to 1.2 ± 0.5 ng/ml in the control group (n=12).

Experiment 2. The influence of androstenone exposures on mouse behavior patterns.

A 30 min exposure of CBA/Lac males to androstenone (160 μL, 0.025%) led to the subsequent changes in a number of behaviors in the odor preference and hole-board tests. Mice of the experimental group (n = 14) did not exhibit the preference for the odor of estrous female, observed in the control group (n = 13), neither by the total time of odor samples investigation nor by the number of approaches toward odor samples (Figure 19.1a, b). In the hole-board test exposure to androstenone promoted a significant increase in the number of rearing episodes (p < 0.05) and a decrease in the duration of grooming (p < 0.01), while the other behavior patterns remained almost unchanged (Table 19.1).

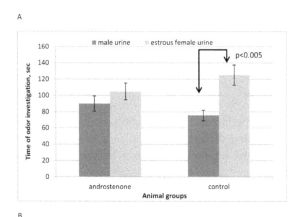

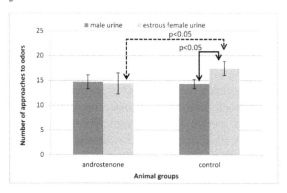

FIGURE 19.1 The influence of androstenone exposures (160 μL, 0.025%, 30 min) on the parameters of CBA/Lac males' behavior in the odor preference test. A – time of investigation of odor samples (sec); —number of approaches to odor samples. The values are presented as group means ± SEM. The observed significant differences between groups are shown in a solid line (Wilcoxon matched pairs test) and in a dashed line (Mann-Whitney U test).

TABLE 19.1   The influence of androstenone exposures on the parameters of behavior of CBA/Lac males in a 10 min hole-board test

Odor exposure groups (160 µL, 30 min)	Head-dips, number	Rearing, number	Grooming, number	Total time of grooming, sec	Defecation (number of boluses)	Urinations, number
Andro-stenone 0.025%, n=14	24,4±1,8	69,9±4,4*	2,86±0,40	9,86±1,92**	2,29±0,42	0,36±0,17
Mineral oil (control odor), n=13	21,7±2,1	50,2±5,0	3,31±0,54	24,6±4,89	2,77±0,52	1,00±0,25

Values are given as group means ± SEM; * $p < 0.05$, ** $p < 0.01$, compared to the control group, Mann–Whitney U test.

19.4   DISCUSSION

The results of the Experiment 1 clearly showed that exposures of CBA/Lac males to the odor of androstenone resulted in a lowered plasma testosterone level if compared to the appropriate control groups. More precisely, we observed testosterone decrease under two experimental conditions: "at rest" in Experiment 1A and in the presence of biologically relevant signal (female estrous odor) in Experiment 1B. Circulating testosterone play a key role in regulation of reproductive and aggressive behaviors in mice. Blood plasma testosterone level oscillates under the control of hypothalamic-hypophyseal-testicles axis. Pulsatile testosterone surges above the baseline are preceded by LH surges (6-8 spontaneous peaks per day) and LH surges are preceded by pulsatile releases of GnRH (Coquelin, Desjardins, 1982). So comparing the experimental group means at the definite moment, we should better describe the androstenone effect rather to postpone or block the nearest testosterone surge than to decrease basal hormone level. Estrous female chemical signals induce reflexive testosterone release in male mice in 20-40 min after stimulus presentation (Macrides et al., 1975, Maruniak et al., 1976, Osadchuk, Naumenko, 1981, Coquelin, Bronson, 1980). This anticipatory release of testosterone occurs due to an "extra" discharge of the endogenous generator of GnRH pulses in hypothalamus and does not differ, either in dynamics or in amplitude, from the spontaneous hormone concentration surges (Nyby, 2008). Thus, the lowered levels of testosterone observed in Experiment 1b could also be explained by the blockage of the upcoming testosterone surge. In our previous studies, we demonstrated that both main and accessory olfactory systems are involved in detection of androstenone in CBA male mice (Voznessenskaya et al, 2010). Following this line of investigation, we suggest that androstenone effects are carried out through the activation of olfactory and/or vomeronasal receptors as far as olfactory structures send neural projections to hypothalamus via amygdala. Meanwhile, we cannot exclude that the hormonal changes could occur due to extracellular route of androstenone delivery to central nervous system from nasal cavity or blood vessels penetration; these routes were described for other steroid substances (Banks et al, 2009, Ducharme et al, 2010). Evidence exists that stress can deactivate the endogenous GnRH pulse generator. But in our study (Experiment 1A) we did not observe related increase in the concentration of the key stress hormone corticosterone. Thus, we conclude that androstenone exposure result in a specific, probably pheromonal effect on blood plasma testosterone in CBA/Lac male mice. We did not observe a significant decrease of testosterone in wild *Mus Musculus* males after androstenone presentations, but only a trend for reduced level of the hormone (Experiment 1B). These results do not contradict data, obtained in CBA/Lac mice, and may attribute to a very low testosterone level (1.2 ± 0.5 ng/ml) observed in the control group of the wild males in the presence of estrus female odor.

A 30 min exposure of CBA/Lac males to the odor of androstenone (160 μL, 0.025%) elicited changes in a number of parameters of behavior in the odor preference and hole-board tests. Preference for the estrous female odor versus male odor, observed in intact CBA/Lac males, can be considered as an expression of the predominance of male sexual motivation. Androstenone exposures disrupted this preference (Experiment 2) that is likely to result in further inhibition of male sexual behavior. In the hole-board test, preliminary androstenone exposures increased rearing, one of the major exploratory behavior indicators. We also registered a decrease in the duration of grooming. Self-grooming, akin scent marking, can be viewed as a form of olfactory communication in mammals (see Ferkin, Leonards, 2010 for review). Self-grooming behavior can be released in response to the odors of conspecifics and may serve to transmit signals of reproductive and physiological status of individual as well as it may facilitate opposite-sex interactions in rodents. On the contrary, the rate of grooming may decrease if a groomer tries to avoid attention of conspecifics. Also we do not exclude that group means for the parameters of behavior in the hole-board test were influenced by the unequal time of exposure to the estrous female odor in the preceding odor preference test. However, the observed behavioral changes in both behavioral tests are consistent with the reduced sexual motivation in response to the exposure to androstenone. The impairments in sexual interest correspond to the marked decreases in plasma testosterone level.

In this and previous (Voznessenskaya et al, 2010) our studies, we collected evidence that androstenone may be involved in chemical communication in mice. In addition to the previously reported intermale aggression promoting properties (Ingersall, Launay, 1986), here we describe a sexual motivation decrease accompanied by testosterone reduction after androstenone exposures in males. Nevertheless, to fulfill criteria for a classical mouse pheromone, androstenone should be present in mouse secretions or excretions. Currently, to the best of our knowledge, there exists no data to solve this issue. If androstenone is somehow excreted or secreted in mice that is likely to be linked to the level of circulating androgens and to signal male dominance. Interestingly, a recent study proposed androstenone involvement in communicating competition and aggression in humans (Lübke, Pause, 2014).

19.5   CONCLUSION

We showed that a volatile steroid androstenone interferes with male mouse reproductive activity by altering circulating sex hormone level and patterns of social behavior. The obtained data may find practical application in the development of safe and nontoxic regulators for the management of pest rodent populations.

ACKNOWLEDGEMENTS

This work was supported by RFBR grant 14-04-01150 and RAS Program "Live Nature".

KEYWORDS

- Mice
- Nontoxic pest population management
- Social behavior
- Steroids
- Testosterone

REFERENCES

1. Banks, W. A.; Morley, J. E.; Niehoff, M. L.; and Mattern, C.; *J. Drug. Target.* **2009**, *17*, 759.
2. BENTON, D.; Biol Psychol, 15, 249 (1982).

3. Cowley, J. J.; JohnsoN, A. L.; Brooksbank, B. W. L.; *Psychoneuroendocrinol.* **1977**, *2*, 159.
4. Coquelin, A.; and Bronson, F.H.; *Endocrinology.* **1980**, *106*(4), 1224.
5. Coquelin, A.; and Desjardins, C.; *Am. J. Physiol.* **1982**, *243*(3), E257.
7. Ducharme, N.; Banks, W. A.; Morley, J. E.; Robinson, S. M.; Niehoff, 7.   M. L.; and Mattern, C.; *Eur J Pharmacol.* **2010**, *641*(2–3), 128.
8. Ferkin, M. H.; Leonards, S. T.; In: Kalueff, A.V. ET AL (eds), Neurobiology of Grooming Behavior. Cambridge University Press; **2010**.
9. Filsinger, E. E.; Braun, J. J.; Monte, W. C.; and Linder, D.E.; *J Comp Psychol.* **1984**, *98*, 220.
10. Filsinger, E.E.; Braun, J. J.; and Monte, W. C.; *Ethol. Sociobiol.* **1985**, *6*, 227.
11. Henttonen, H.; Oksanen, T.; Jortikka, A.; and Haukisalmi, V.; *Oikos.* **1987**, *50*(3), 353.
12. Ingersoll, D. W.; and Launay, J.; *Physiol. Behav.* **1986**, *36*(2), 263.
13. KIRK-SMITH, M. D.; BOOTH, M. A.; CARROL, D.; DAVIES, P.; *Res. Commun. Psychol. Res. Psychiat. Behav.* **1978**, *3*, 379.
14. Levai, O.; Feistel, T.; Breer, H.; and Strotmann, J.; *J Comp Neurol.* **2006**, *498*(4), 476.
15. Lübke, K.; and Pause, B. M.; *Physiol. Behav.* **2014**, *128*, 52.
16. Macrides, F.; Bartke, A.; and Dalterio, S.; *Science.* **1975**, *189*(4208), 1104.
17. Maruniak, J. A.; and Bronson, F. H.; *Endocrinology.* **1976**, *99*(4), 963.
18. Nodari, F., Hsu, F. F.; Fu, X.; Holekamp, T. F.; Kao, L. F.; Turk, J.; and Holy, T.E.; *J. Neurosci.* **2008**, *28*(25), 6407.
19. Osadchuk, A. V.; and Naumenko, E. V.; *Dokl. Akad. Nauk* SSSR. **1981**, *358*(3), 746.
20. Pause, B.M.: *Physiol Behav.* **2004**, *83*, 21.
21. Reed, H. B. C.; Melrose, D. R.;and Patterson, R. L. S.; *Br. Vet. J.* **1974**, *130*, 61.
22. Rylnikov, V. A.; Savinetskaya, L. E.; and Voznessenskaya, V. V.; *Soviet. J. Ecol.* **1992**, *23*(1), 46.
23. Sam, M.; Vora, S.; Malnic, B.; MA, W.; Novotny, M.V.; and Buck, L. B. *Nature.* **2001** *412*, 142.
24. Van Toller, C.; Kirk-Smith, M.; Wood, N.; Lombard, J.; Dodd, G. H.; *Biol. Psychol.* **1983**, *16*, 85.
25. Voznessenskaya, V. V.; In: C. Musignat-Caretta (Ed), Neurobiology of Chemical Communication (Frontiers in Neuroscience Book Series). CRC Press, **2014**, 389.
26. Voznessenskaya, V. V.; Klyuchnikova, M. A.; and Wysocki, C. J.; *Curr. Zool.* **2010**, *56*(6), 813.
27. Voznessenskaya, V. V.; and T.V.Malanina: In: G.E. Zaikov (Ed.): News in Chemistry, Biochemistry and Biotechnology: State of the Art and Prospects of Development. Nova Science Publishers: NY, **2014**; *59*.
28. Voznessenskaya, V. V.; Naidenko, S.V.; Clark, L.; Pavlov, D.S.; In: G.E. Zaikov (Ed.): Biotechnology and the Environment Including Biogeotechnology. Nova Science Publishers: NY, **2004**; 59.
29. Voznessenskaya, V. V.; Parfyonova, V. M.; W AND Ysocki, C. J. *Adv Biosci.* **1995**, *93*, 399.
30. Voznessenskaya, V. V.; and Poletaeva, I. I.; *Zhurnal Vysshei Nervnoi Deyatelnosti imeni I P Pavlova.* **1987**, 37(1), 174–176.
31. Voznessenskaya, V. V.; and Wysocki, C. J.; *Chem. Sense.* **1994**, *19*, 569.
32. Voznessenskaya, V. V.; Wysocki, C. J.; Chukhrai, E. S.; Poltorack, O. M.; Atyaksheva, L. F.; In: R.E. Johnston et al (eds), Advances in Chemical Signals in Vertebrates, Kluwer: NY, **1999**; 563.
33. Wysocki, C. J.; Preti, G.; http: //senseofsmell.org/papers/Human_Pheromones_Final%20 7-15-09.pdf (**2009**).

CHAPTER 20

OLFACTORY FUNCTION AS MARKER OF NEURODEGENERATIVE DISORDERS: TESTS FOR OLFACTORY ASSESSMENT AND ITS APPLICABILITY FOR RUSSIAN POPULATION

V. V. VOZNESSENSKAYA[1], A. E. VOZNESENSKAYA[2], and M. A. KLYUCHNIKOVA[1], E. I. RODIONOVA[2]

[1]A. N. Severtzov Institute of Ecology and Evolution, 33 Leninski prospect, Moscow, 119071, Russia; E-mail: veravoznessenskaya@gmail.com

[2]A. A. Kharkevich Institute for Information Transmission, 19 B. Karetny, Moscow, 127994, Russia

CONTENTS

20.1   INTRODUCTION

Olfactory system disorders received much clinical and experimental attention within last two decades as several severe neurological conditions have been associated with peripheral or central deficits of the olfactory system. Some neurodegenerative disorders such as Alzheimer disease (AD) and Parkinson disease (PD) are characterized by early olfactory loss. A number of studies demonstrated the impairments of olfactory memory at early stages of AD [8; 5]. Inability to identify odors in AD patients is secondary relative to inability to detect odors in principle. Olfactory disorders in AD are bilateral and progress as disease is developing [13], [9], [10]. About 95 percent of PD patients show olfactory deficit at preclinical stages of disease. Unlike AD, there is no apparent longitudinal progression in olfactory dysfunction as occurs in other elements of the disease process [3]. In 2006 American Academy of Neurology recommended evaluation of olfactory function for differential diagnostics of PD [15]. Standardized olfactory tests are widely used in clinical practice in the United States for differential diagnostics of neurological disorders. A number of tests based on odor identification are currently applied around a world. Olfactory deficit in AD could be detected by a wide range of olfactory tests, including tests for odor identification, detection threshold sensitivity, discrimination and memory. Nevertheless smell identification tests when respondents requested to select right odor name out of multiple alternates turned out to be the most sensitive. High sensitivity of such tests are explained by pathophysiology of AD development. Development of AD starts from the structures of olfactory system. Entorinal and trans-entorinal brain area, hippocampus and amigdala are the first sites where neurite plates appear (0 stage of AD in accordance with Braque classification). Thus, structures which are critical for olfactory sensitivity and olfactory memory, suffer at the very beginning of AD. Early diagnosis of neurodegenerative diseases such as AD is important for increasing the efficacy of medications that may thwart the progression of the symptoms.

The University of Pennsylvania Smell Identification Test (UPSIT) is a scratch and sniff test used in USA since 1984 [2]. Test is accompanied by detailed instructions of olfactory score calculation. The benefits of this test are high reliability, easy assessment, and high repeatability of the results ($r = 0.90$–0.92). Despite its importance no test assessing olfactory function is developed so far in Russian Federation. The aim of this study was to access potential applicability of UPSIT for the assessment of olfactory function in Central Russia population and develop Russian Smell Identification Test. Multiple choice olfactory identification tests quickly reveal a sensory dysfunction. Though, the results of these tests are affected by cultural backgrounds. It is very important that odorants used, as well as the odor names suggested as possible answers belong to the cultural background of the tested population. As the UPSIT was developed in US it contains some samples of product odors introduced in Russian culture within only last 15 years. Detection of unidentifiable and unfamiliar odors within testing samples is one of the important steps of applicability assessment.

20.2 TEST SUBJECTS AND PROCEDURE

Test subjects were healthy volunteers of 18–79 years of age, males and females, with no apparent olfactory problems: Moscow residents (n = 145) as well as people from rural area (n = 107). More than 70 percent of subjects reported their ethnicity as "Russian." Based on age we subdivided all test subjects into four groups. The profound influence of age on the olfactory function is well-known from the literature. This observation also supported by our own data (Figure 20.1). Under age of 18 olfactory functioning may be seriously affected by unstable basal levels of steroid hormones. Significant drop of olfactory sensitivity noted for people over 60–65 years old. Taking this into consideration, we used group of test subjects of 18–59 years old and group—over 60 years old. In control group (18–35 years old) NO concentrations lack in nasal cavity was monitored (Logan Research-2149) to exclude individuals with allergic inflammations.

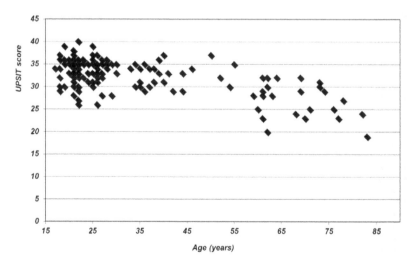

FIGURE 20.1 The influence of age on the performance of respondents from Russian Federation in the University of Pennsylvania Smell Identification Test (n = 181).

We used methodology of administration recommended by manufacturer of University of Pennsylvania Smell Identification Test™, UPSIT (Sensonics, Inc, Haddo Heights, NJ). The response alternatives of UPSIT were translated into Russian with changing some odorant labels: "wintergreen" to "methylsalicylate containing products (Orbit Wintergreen/warming ointments)" and "smoke" to "smoked meat products." The response alternatives were provided in written form printed on separate sheets of paper. The instructions were read to each subject by an experimenter who monitored the test session to ensure that the correct procedure was followed. Additionally each test subject was asked to fill out questionnaire. Survey included such questions like smoking, hormonal treatments, level of education, etc. Additional group of test subjects (n = 38) was asked to rate intensity of odors (low, moderate, strong) or report if they couldn't detect the odorant.

After UPSIT administration respondents were given a list of all 51 response alternatives of the UPSIT (including absent odors such as skunk, honey, etc.) and asked to mark completely unfamiliar odors. Afterwards the subjects were provided with correct answers and asked if the latter were consistent with their conception for particular odor.

We monitored level of stress of test subjects during the testing procedure based on saliva cortisole. Saliva cortisole was measured by ELISA technique using EIA kits of reagents (EIA SLV4635, DRG, USA).

20.3   RESULTS AND DISCUSSION

Average scores of correct answers out of 40 odor samples was $34,73 \pm 2,29$ in Moscow (n=99) and 33.08 ± 3.12 in rural area (Palekh, n = 60) for individuals from 18 to 59 years old which is comparable with scores reported for North America, Italy and Greece [2], [11]; [4]. For individuals over 60 years old we received the following results: $28,22 \pm 3,35$ (Moscow, n = 46) and $27,36 \pm 3,70$ (n = 47, Palekh). No statistically significant difference between men and women as well as between Moscow and Palekh residents in test performance was detected (Table 20.1). As a control a group of people (18–35 years old) with known medical history was used. Lack of allergical inflammation was confirmed by NO level measurement in the nasal cavity of these subjects. Average UPSIT score for control group was 33.08 ± 2.87 (n = 17) which also did not differ from age matched experimental groups. We did not observe the influence of smoking on performance in UPSIT. For age matched groups (18–35 years old, n = 134) UPSIT score appeared as 34.11 ± 0.34 and 33.61 ± 0.30. This is in good agreement with Silveira-Moriyama et al., who also did not observe score differences in UPSIT between smokers and nonsmokers [12]. Level of education (high school graduates or BS/MS holders) did not affect UPSIT score for the same age group which is very important for applicability of the test.

TABLE 20.1 Performance of respondents from Russian Federation in the University of Pennsylvania Smell Identification Test (n = 252).

Experimental group (age)	N	UPSIT score (±SD)
Moscow (18-59)	99	34,73±2,29
Rural area (18-59)	60	33.08 ± 3.12
Moscow (≥60)	46	28,22±3,35
Rural area (≥60)	47	27,36±3,70

In North America for healthy people under 60 years old average score in UPSIT ranges from 34 to 40. The reason for lower scores in RF population may be cultural differences. Many smells are culturally specific. It means that an olfactory identification test must take into account the cultural background of tested individuals. Another reason could be environmental influences which may affect olfactory thresholds [18], [17], [14]. To assess these factors we analyzed identifiability of odors included into UPSIT. In original test 75 percent ofidentifiability was used as criterion for odor inclusion into UPSIT. We used the very same threshold adapting UPSIT for RF population [16].

More than 95 percent of respondents both in Moscow and rural area correctly named 17 odor samples (licorice, rose, cherry, banana, coconut, turpentine, pine, gasoline, peach, natural gas, mint, wintergreen, lemon, leather, gingerbread, onion, smoke). Though, less than 50 percent of test subjects identified five odors: "lime," "fruit punch," "lilac," "cheddar," and "grass." The most frequent alternate for "fruit punch" was "soap" (95 percent of respondents) which allowed us to rename the odorant to "soap" not excluding from the test. For other four odor samples respondents suggested different alternates. As far as these four odor samples turned to be unrecognizable for majority of subjects we excluded them from the test. Intensity rating of the odor samples revealed that only 10 percent were perceived as "weak." Interestingly, that among these samples occurred four which we already excluded from the test as "unidentifiable." Our data indicate applicability of short version (32 odor samples) of UPSIT to assess olfactory function of RF population. We also adapted numerical key (Table 20.2). Score 17–22 indicates high risk for the development of neuro-degenerative disorders for males and females. Shorten test is quite often used in different countries to screen olfactory function [7]. Nevertheless full size test is preferable. Inability to identify certain odors may be more reliable symptom of PD. For instance inability to identify "pizza" odor or "wintergreen" odor is more serious sign of developing PD then inability to identify odor of "onion," "fruit punch," or "turpentine" [6]. After consulting with UPSIT manufacturer we performed additional experiments to replace excluded odor samples. We were suggested with 7 odor samples as potential substitutes. Results presented in Table 20.3. "Garlic," "grapefruit," "rubber tire," coffee" odor samples were selected to replace unidentifiable odorants.

TABLE 20.2   Scoring for Russian version of University of Pennsylvania Smell Identification Test

Olfactory Function	Gender	Цифровое значение
Norma	female	32-36
Norma	male	31-36
Light anosmia	female	28-31
Light anosmia	male	27-30
Moderate anosmia	female	23-27
Moderate anosmia	male	23-26
Severe anosmia	female, male	17-22
Total anosmia	female, male	5-16

TABLE 20.3   Identifiability of odor samples suggested for inclusion in the Russian version of University of Pennsylvania Smell Identification Test (n = 86).

Odor name	18–35 years old /Moscow, n = 21	52–82 years old /Moscow, n = 24	18–35 years old /Palekh, n = 20	55–82 years old /Palekh n = 21
Apple	33.3% (7)	66.7% (16)	80% (16)	76.2% (16)
Raspberry	57.1% (12)	91.7% (21)	80% (16)	85.7% (18)
Garlic	100% (21)	100% (24)	100% (20)	100% (21)
Baby powder	90.5% (18)	100% (24)	75% (15)	85.7% (18)
Grapefruit	95.2% (20)	95.8% (23)	100% (20)	95.2% (20)
Rubber tire	76.2% (16)	87.5% (21)	95% (19)	95.2% (20)
Coffee	95.2% (20)	100 (24)	100% (20)	100% (21)

20.4 CONCLUSIONS

Russian version of Smell Identification Test (RSIT) has been developed based on University of Pennsylvania Smell Identification test. The RSIT is suitable for clinical use as well as for population screening. Introduction of the test into medical practice would facilitate preclinical diagnostics of neurodegenerative disorders in Russia.

ACKNOWLEDGMENTS

This research was supported by Russian Academy of Sciences, Program "Basic Sciences—to Medicine" to VVV

KEYWORDS

- **Humans**
- **Neurodegenerative disorders**
- **Odor perception**
- **Olfactory function**
- **University of Pennsylvania Smell Identification test**

REFERENCES

1. Djordjevic, J.; Jones-Gotman, M.; De Sousa, K.; and Chertkow, H.; *Neurobiol. Aging.* **2008**, *29*(5), 693–706.
2. Doty, R. L.; Shaman, P.; and Dann, M.; *Physiol. Behav.* **1984**. *32*, 489–502.
3. Doty, R. L.; Deems, D. A.; and Stellar, S.; *Neurology.* **1988**, *38*, 1237–1244.
4. Economou, A. ; *Arch. Gerontol. Geriatr.* **2003**. *37*(2), 119–130.
5. Gilbert, P. E.; Barr, P. J.; Murphy, C.; *J. Int. Neuropsychol. Soc.* **2004**, *10*(6), 835–842.
6. Hawkes, C. H.; and Shephard, B. C.; *Ann N Y Acad Sci.* **1998**, 855, 608–615.
7. Kobayashi, M.; Reiter, E. R. Dinardo, L. J.; and Costanzo, R. M.; *Arch Otolaryngol. Head. Neck. Surg.* **2007**. *133*(4), 331e.
8. Lehrner, J. P.; Brucke. T.; Dal-Bianco, P.; Gatterer, G.; Kryspin-Exner, I.; *Chem. Senses.* **1997**, *22*(1), 105–110.
9. Murphy, C.; Gilmore, M. M.; Seery, C. S.; Salmon, D. P.; and Lasker, B. R.; *Neurobiol Aging*, *11*(4), **1990**, 465–469.
10. Nordin, S.; Almkvist, O.; Berlund, B.; and Wahlund, L. O.; *Arch Neurol.* **1997**, *54*(8), 993–998.
11. ParolA, S.; and Liberini, P.; *Italian. J. Neurol. Sci.* **1999**, *20*(5), 287–296.
12. Silveira-Moriyama, L.; Carvalho, Mde J.; Katzenschlager, R.; Petrie, A.; Ranvaud, R.; Barbosa, E. R.; *Mov Disord.* *23*(16), **2008**, 2328e34.
13. Serby., M.; Corwin, J.; Conrad, P.; Rotrosen, J.; *Am. J. Psych.***1985**, *142*(6), 781–782.

14. Sokolov, V. E.; Voznesenskaya, V. V.; Parfenova, V. M.; and Wysocki, C. J.; *Doklady Akademii Nauk.* **1996,** *347*(6), 843–846.
15. Suchowersky, O.; Reich, S.; Perlmutter., J.; Zesiewicz, T.; Gronseth, G.; Weiner, W. J.; *Neurology.* **2006,** *66*(7), 968–975.
16. Voznesenskaya, A.; Klyuchnikova, M.; Rodionova, E.; Voznessenskaya, V.; *Chem. Senses,* **2011,** *36*(1), E74–E75.
17. Voznessenskaya, V. V.; ParfyonovA, V. M.; and Wysocki, C. J.; *Adv. Biosci.* **1995,** *93*, 399.
18. Wysocki, C.J.; Dorries, K.; and Beauchamp, G. K.; *Proc.Natl.Acad.Sci.USA.* **1989,** *86*, 7976–7978 ()

CHAPTER 21

MODIFICATION OF RECEPTOR STATUS IN GROUPS OF PROLIFERATIVE ACTIVITY OF BREAST CARCINOMAS

A. A. BRILLIANT, S. V. SAZONOV, and Y. M. ZASADKEVICH

GBUZSO Institute of Medical Cell Technologies, 620036 Yekaterinburg, Soboleva str., 25; E-mail: zasadkevich@celltechnologies.ru

CONTENTS

21.1   INTRODUCTION

It is known, that some prognostic and predictive factors should be considered to solve the problem of the therapy of breast carcinomas. A prognostic importance of the cell proliferation index (Ki67) is significant for those tumors, in which it is difficult to predict clinical course only with histological characteristics [1]. The cell proliferation index is an independent predictor of the general survival as well as the disease relapse in breast carcinoma patients [2]. Additionally, univariate and multivariate analyses show that the cell proliferation index correlates with unfavorable clinical outcome [1, 3, 4].

During the breast carcinoma tissue examination, determination of steroid hormones receptors such as Estrogen (ER) and Progesterone (PR) is almost always used. It plays a key role in the correct assignment of hormonal therapy [5]. It is known that in three groups (ER+ PR+), (ER- PR+) and (ER+ PR-) resistance to hormone therapy (tamoxifen) is determined more often in the third group. It can be explained by the fact that in patients with "ER+ PR-" status the level of Her-2/neu and EGRF is higher [6].

Her-2/neu is a protooncogene, which encodes human epidermal growth factor receptor 2 (c-erb-2), related to tyrosine kinase group. Hyperexpression of this oncogene is observed in 25-30% of breast cancer cases and associates with a poor prognosis in both with presence of metastases or without them [7].

The above listed tumor biomarkers such as Ki67, ER, PR, Her-2/neu are recommended for widespread clinical use nowadays [8]. The combination of them gives necessary information about receptor status of a tumor. However, dynamics of processes in a tumor remains still unclear. Study of dependence of receptor status from proliferation complicates with heterogeneity of the explored tumors.

21.2   MATERIALS AND METHODS

Selected cases were analyzed by histological, immunohistochemical, morphometric and statistical methods. 406 cases of breast carcinomas were studied. Material for research was supplied by Sverdlovsk regional cancer center and Municipal Mammological Centre of Clinical hospital №40, Ekaterinburg, Russia. For immunohistochemical method glasses covered by polysine POLYSINE SLIDES (Thermo scientific, Germany) were used. Her-2/neu expression at tumor cells was detected by polyclonal antibodies Polyclonal Rabbit Anti-Human c-erb-2 Oncoprotein (DAKO, Denmark), Estrogen and progesterone receptors expression was detected by monoclonal antibodies Monoclonal Mouse Anti-Human Estrogen Receptor, Monoclonal Mouse Anti-Human Progesterone Receptor (DAKO, Denmark). Level of proliferative activity was studied by evaluation of cell proliferation biomarker (Ki67) expression. For detection of the cell proliferation index antibodies Mouse Anti-Human KI-67 Antigen (DAKO, Denmark) were used. Proliferative activity of the investigated tumor can be evaluated by percentage ratio of stained nuclei of breast carcinoma cells to unstained. Immunohistochemical tests were made in autostainer "DAKO" (Denmark) with use of Dako Wash Buffer, visualization system Dako EnVision+Dual Link System-HRP, chromogen Dako Liquid DAB+ Substrate Chromogen System (DAKO, Denmark). Test evaluation was made with the robitic microscope "Zeiss Ymager M" (Germany). Membrane expression of Her-2/neu in tumor cells was evaluated on the scale from 0 to 3+ [9]. Level of estrogen and progesterone receptors expression was detected on the scale from 0 to 8+ [10].

21.3   RESULTS

After processing of 406 cases 3 groups of patients accordingly to Ki-67 protein expression were formed. The first group contained 248 cases (61% of the explored cases) in which Ki67 expression was lower than 10 percent inclusively. The second group contained 82 cases (20% of explored cases), in which Ki67 expression was higher than 10% but lower than 30% inclusively. The third group contained 76 cases (19% of explored cases), in which Ki67 receptors were detected in more than 30 percent of tumor cells (Figure 21.1). Division of cases of patients with breast carcinoma according to percentage of tumor cells expressed Ki67 is shown at the Figure 21.2.

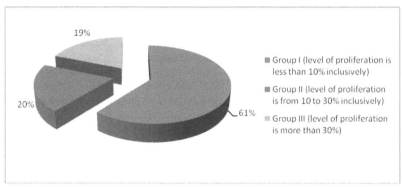

FIGURE 21.1   Distribution of patients with breast carcinoma according to a percentage ratio of cell expressing Ki-67.

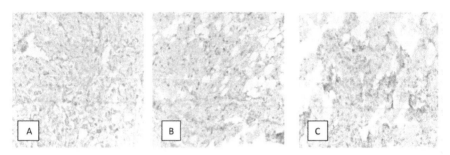

FIGURE 21.2   Breast carcinoma with level of proliferation A-10% of tumor cells, B-30% of tumor cells, C-50% of tumor cells. X100. Staining: IHC reaction HRP/DAB, additional staining with Mayer's hematoxylin.

Thus, a majority of breast carcinomas has a low level of proliferation that is less than 10% of tumor cells express Ki67 protein receptors.

Additional research, allowing evaluating features of receptor profile of a tumor according to its proliferative activity, was conducted (Table 21.1). After comparative analysis of 406 cases it was found that group I (degree of proliferation is less than 10%) does not have any significant difference in receptor status in comparison with group II (with degree of proliferation 10–30%). In turn, group II and group III (degree of proliferation is more than 30%) have some differences in their receptor profile. With increasing of proliferative activity of carcinoma cells expression of steroid hormones (Estrogen receptor, Progesterone receptor) is decreasing and Her-2/neu expression is increasing (Table 21.1).

TABLE 21.1   Receptor status of breast carcinoma in groups of Ki-67 expression

Group of Ki-67 expression	Average definition of receptor status	
Group I – cases with low level of proliferation (less than 10% inclusively) n=248 (61%)	Estrogen receptor	2.5±0.2; p(1-2) >0.05
	Progesterone receptor	2.2±0.2; p(1-2) >0.05
	HER-2 /neu	1.9±0.8; p(1-2) >0.05
Group II – cases with medium level of proliferation (more than 10%, less than 30% inclusively) n=82 (20%)	Estrogen receptor	2.3±0.2; p(2-3)<0.05
	Progesterone receptor	2.0±0.2; p(2-3)<0.05
	HER-2 /neu	2.0±0.9; p(2-3) <0.05
Group III – cases with low level of proliferation (more than 30%) n=76 (19%)	Estrogen receptor	1.5±0.3; p(3-1) <0.05
	Progesterone receptor	1.3±0.2; p(3-1) <0.05
	HER-2 /neu	2.8±1.5; p(3-1) <0.05

Due to heterogeneity of analyzed carcinomas, included in the groups of proliferation, it was decided to conduct a study of a degree of proliferation in combination with presence or absence of expression of steroid receptors and Her-2/neu. It was detected that only 13 cases (3%) out of all 406 cases had a positive expression of Estrogen, Progesterone and Her-2/neu at the same time. After analyzing of the division of these cases for groups of proliferative activity, we found a significant increase of steroid hormones expression from the first to the second group and decrease to the third group of proliferation (Table 21.2). We did not find any significant difference in Her-2/neu expression in the groups of proliferation. Thus, in this group another dependence of change of level of receptor expression in changing of proliferative processes was found in comparison with the general result.

TABLE 21.2   Receptor status of carcinoma in groups of Ki-67 expression with positive steroid hormones receptors and Her-2/neu expression

Group of Ki-67 expression	Average definition of receptor status	
Group I – cases with low level of proliferation (less than 10% inclusively) n=6 (46%)	Estrogen receptor	3.6±0.8; p(1-2) <0.05
	Progesterone receptor	4.0±1.3; p(1-2)<0.05
	HER-2 /neu	2.3±0.4; p(1-2) >0.05
Group II – cases with medium level of proliferation (more than 10%, less than 30% inclusively) n=4 (30%)	Estrogen receptor	5.3±2.3; p(2-3)<0.05
	Progesterone receptor	5.0±1.4; p(2-3)<0.05
	HER-2 /neu	2.2±0.1; p(2-3)>0.05
Group III– cases with low level of proliferation (more than 30%) n=3 (23%)	Estrogen receptor	3.9±1.3; p(3-1) >0.05
	Progesterone receptor	4.1±2.3; p(3-1) >0.05
	HER-2 /neu	2.1±0.1; p(3-1) >0.05

The next step of our research was to study a dependence of level of estrogen receptor expression from level of activity of proliferative processes in the group of carcinomas positive only to estrogen receptor and negative to progesterone receptor and Her-2/neu. 60 appropriate cases (15% from all carcinomas) were explored. The comparative analysis showed that the level of estrogen receptor expression in the first group of proliferative activity was significantly lower, than in the second and the third groups (Table 21.3). Therefore when the level of proliferation of tumors increases, Estrogen receptor expression grows as well. When levels of Progesterone receptor and Her-2/neu expression in the groups of research were compared, significant difference was not found. It is worth noting that the number of cases with high level of proliferation and positive Esrtrogen receptor expression is less in 3 times than cases with low proliferative activity.

TABLE 21.3 Receptor status of carcinoma in groups of Ki-67 expression with positive expression of Estrogen receptor

Group of Ki-67 expression	Average definition of receptor status	
Group I – cases with low level of proliferation (less than 10% inclusively) n=27 (45%)	Estrogen receptor	3.5 ± 0.2; p(1-2) <0.05
	Progesterone receptor	0.01 ± 0.01; p(1-2) >0.05
	HER-2 /neu	0.3 ± 0.1; p(1-2) >0.05
Group II – cases with medium level of proliferation (more than 10%, less than 30% inclusively) n=24 (40%)	Estrogen receptor	4.5 ± 0.3; p(2-3) >0.05
	Progesterone receptor	0.01 ± 0.01; p(2-3) >0.05
	HER-2 /neu	0.3 ± 0.1; p(2-3)>0.05
Group III – cases with low level of proliferation (more than 30%) n=9 (15%)	Estrogen receptor	4.8 ± 0.4; p(3-1) <0.05
	Progesterone receptor	4.1 ± 2.3; p(3-1) >0.05
	HER-2 /neu	0.60 ± 0.04; p(3-1) >0.05

During the research of dependence of level of Progesterone receptor expression from level of proliferative activity, positive only to Progesterone receptor and negative to Estrogen receptor and Her-2/neu, we found 32 cases (8% from all the breast carcinomas) appropriate under these criteria. We did not find any significant difference of expression of these receptors in the groups of proliferative activity; hence a dependence of progesterone receptor expression from proliferative activity of the tumor was not detected (Table 21.4). Her-2/neu expression in the groups of proliferative activity does not have any significant difference. Estrogen receptor expression does not change with the increase of the level of proliferation of the tumor. Number of cases of the third group (with high level of proliferation) is 3 and 6 times lower than the number of cases in the second and the first groups of Ki67 expression, respectively.

TABLE 21.4 Receptor status of carcinoma in groups of Ki-67 expression with positive expression of Progesterone receptor

Group of Ki-67 expression	Average definition of receptor status	
Group I – cases with low level of proliferation (less than 10% inclusively) **n=19 (60%)**	**Estrogen receptor**	**0.02 ± 0.02; p(1-2)** >0.05
	Progesterone receptor	**4.3 ± 0.3; p(1-2)** $>0,05$
	HER-2 /neu	**0.01 ± 0.00; p(1-2)** >0.05
Group II – cases with medium level of proliferation (more than 10%, less than 30% inclusively) **n=10 (30%)**	**Estrogen receptor**	**0.04 ± 0.04; p(2-3)** >0.05
	Progesterone receptor	**4.3 ± 0.4; p(2-3)** >0.05
	HER-2 /neu	**0.09 ± 0.06 p(2-3)>0.05**
Group III – cases with low level of proliferation (more than 30%) **n=3 (10%)**	**Estrogen receptor**	**0.0 ± 0.0; p(3-1)** >0.05
	Progesterone receptor	**3.8 ± 0.8; p(3-1)** >0.05
	HER-2 /neu	**0.6 ± 0.2; p(3-1)** >0.05

From all 406 cases of infiltrative carcinoma 129 (32%) express Estrogen receptor, Progesterone receptor and negative to Her-2/neu at the same time. During the research of dependence of level of steroid hormones receptors expression on the level of proliferative activity of breast carcinoma in this group no significant difference in the change of proliferative activity was not found (Table 21.5). A small number of studied carcinomas related to the first group of proliferative activity. The more proliferative processes in a tumor are the less number of cases with positive estrogen and progesterone receptors are. Number of cases decreases with the increase of proliferative activity from the first to the third group in four times.

TABLE 21.5   Receptor status of carcinoma in groups of Ki-67 expression with positive expression of Estrogen and Progesterone receptors

Group of Ki-67 expression	Average definition of receptor status	
Group I – cases with low level of proliferation (less than 10% inclusively) n=62 (48%)	Estrogen receptor	4.3±0.1 p(1-2) >0.05
	Progesterone receptor	4.8±0.2; p(1-2) >0.05
	HER-2 /neu	0.24±0.03; p(1-2) >0.05
Group II – cases with medium level of proliferation (more than 10%, less than 30% inclusively) n=50 (39%)	Estrogen receptor	4.8±0.2; p(2-3) >0.05
	Progesterone receptor	4.9±0.2; p(2-3) >0.05
	HER-2 /neu	0.24±0.04; p(2-3)>0.05
Group III – cases with low level of proliferation (more than 30%) n=17 (13%)	Estrogen receptor	5.0±0.3; p(3-1) >0.05
	Progesterone receptor	5.3±0.3; p(3-1) >0.05
	HER-2 /neu	0.2±0.1; p(3-1) >0,05

41 cases (10 percent from all carcinomas) were included in the group of carcinomas, expressed Her-2/neu receptor (2+ and 3+ cases) and which were negative to Estrogen and Progesterone receptors. A half of all the cases was included in the second group of Ki67 expression and had a medium level of proliferation. We did not find any significant difference in Her-2/neu expression in the groups of proliferation thus dependence of Her-2/neu from proliferative activity of a tumor was not detected (Table 21.6).

TABLE 21.6 Receptor status of carcinoma in groups of Ki-67 expression with positive expression of Her-2/neu receptor

Group of Ki-67 expression	Average definition of receptor status	
Group I – cases with low level of proliferation (less than 10% inclusively) n=9 (23%)	HER-2 /neu	2.4±0.1; p(1-2) >0.05
Group II – cases with medium level of proliferation (more than 10%, less than 30% inclusively) n=23 (54%)	HER-2 /neu	2.5±0.1; p(2-3)>0.05
Group III – cases with low level of proliferation (more than 30%) n=9 (23%)	HER-2 /neu	2.3±0.1; p(3-1) >0.05

We conducted an additional research allowing studying distribution of the cases with negative expression of Estrogen, Progesterone receptors and Her-2/neu to the groups of proliferation. In the result of studying of 64 cases (16%) it was found that 25% and 28% respectively related to the groups with low and medium levels of proliferation. The majority of cases were attributed to the third group (with high level of proliferation—44 percent from all the cases).

21.4 CONCLUSION

After analyzing the received data, it can be concluded that the general study, which does not consider heterogeneity of properties of the tumors included in the study, could be used for the evaluation of receptor profile of breast carcinomas. Thereby, distribution to the groups of proliferation of all the cases showed that steroid hormones receptors expression decreased and Her-2/neu expression increased with increasing of proliferative activity of breast carcinoma cells. Differentiated approach at the research of receptor status in the groups of tumor proliferation showed that number of tumors with high proliferative activity grows with the increase of Estrogen receptor expression. We did not find any significant difference in Progesterone receptor expression in change of level of proliferation in a tumor. Moreover, a majority of positive at Estrogen and Progesterone receptors expression cases related to the group with low level of proliferation. It was not also find any significant difference at Her-2/neu expression in the different groups of proliferative activity. The group with the medium level of proliferation included twice more cases of carcinomas with positive expression of Her-2/neu.

KEYWORDS

- **Breast carcinoma**
- **Immunohistochemistry**
- **Proliferation**
- **Receptor status**

REFERENCES

1. Scholzenm, T.; The Ki-67 protein: from the known and the unknown. *J. Cell. Physiol.* **2000**. *182*, 311.
2. Penault-Llorca, F.; Cayre, A.; and Bouchet Mishellany, F. et al. Induction chemotherapy for breast carcinoma: predictive markers and relation with outcome. *Int J Oncol,* **2003**, 25–1319.
3. Gonzalez-Vela, M. C.; Garijo, M. F.; Fernandez, F.; and Val-Bernal, J. F.; MIB1 proliferation index in breast infiltrating carcinoma: comparison with other proliferative markers and association with new biological prognostic factors. *Histol. Histopathol.* **2001**, *16*, 399–406.
4. Jones, S.; Clark, G.; and Koleszar, S. et al. Low proliferative rate of invasive nodenegative breast cancer Predicts for a favorable outcome: a prospective evaluation of 669 patients. *Breast Cancer.* **2001**. 1, 310–314.
5. Sazonov, S. V.; and Leontiev, S. L.; Creation of the system of revision of immunohistochemical research in breast cancer diagnostics. *Vest. Ural. Acad. Sci.* **2012**, *1*(38), 18–23.
6. Garin, A. M.; Endocrine therapy and hormone dependant tumors. M. Tver: Triada, **2005**; 240 p.
7. Joerger, M.; Thürlimann, B.; and Huober, J.; Small HER2-positive, node-negative breast cancer: who should receive systemic adjuvant treatment?. *Ann Oncol.* **2011** *22*(1), 17–23.
8. Frank G.; Zavalishina, L.; Andreeva, J.; Matsionis, A.; and Sazonov, S.; HER2 testing in Russia: The results of the 10 years of experience. ASCO/CAP recommended score system. *Wirchows Archiv,* **2012**, *461* (Suppl 1), 241–242.
9. Bilous, M., Dowsett, M., Isola, J., Lebeau, A., and Moreno, A., et al. Current perspectives on HER2 testing: a review of national testing guidelines. *Mol Pathol.* **2003**, *16*, 82–173.
10. Elledge, R. M.; Green, S.; Pugh, R.; Allred, D. C.; Clark, G. M.; Hill, J. et al. Estrogen receptor (ER) and Progesterone receptor (PgR), by ligand-binding assay compared with ER, PgR and pS2, by immunohistochemystry in predicting response to tamoxifen in metastatic breast cancer: a Southwest Oncology Group Study. *Int. J. Cancer.* **2000**. 89, 7–11.

CHAPTER 22

A RESEARCH NOTE ON RESUSCITATION OF VIABLE BUT NONCULTURABLE PROBIOTIC BACTERIA

YU. D. PAKHOMOV, L. P. BLINKOVA, O. V. DMITRIEVA, O. S. BERDYUGINA, and N. N. SKORLUPKINA

FGBU "I.I. Mechnikov Research Institute for Vaccines and Sera" RAMS. Moscow 105064 Maliy Kazenniy per. 5a. E-mail: labpitsred@yandex.ru

CONTENTS

22.1   INTRODUCTION

Existence of viable but nonculturable bacteria and their resuscitation into vegetative state are the major problems in modern microbiology.Nonculturable cells
are formed as a response to a wide variety of stressful factors and their combinations. It is particularly important for food producers and manufacturers of
probiotic preparations, since bacteria may loseculturability as a reaction to sani-
tation procedures, lyophilization or during fermentation processes in ripening
foodstuffs [3, 9, 4, 1]. Main task is to search for resuscitating agents that con-
vertnonculturable cells into active forms.According to literary data such agents
include fetal serum, vitamin K, yeast cells, live cells of higher organisms, etc.
[7, 12, 11, 10]. Our aim was to study several resuscitating factors were tested on
nonculturable cells of *Lactococcuslactis* and a number of lyophilized probiotics.

22.2   MATERIALS AND METHODS

Strains and media:In this study we used following microorganisms: lyophilized
probiotic cultures of *E. coli* M-17, *Bifidobacteriumbifidum, Lactobacillus
acidophilus* and three nisin producing strains of *Lactococcuslactis ssp.
lactis*MSU, 729 and F-116 that were incubated in carbohydrate starvation
conditions [8] and contained more than 99.9 percent nonculturablecells.

For growing microorganisms following media were used: tryptic soy
medium for *E. coli*, Elliker medium for *L. lactis,B. bifidum*and *Lactobacillus
acidophilus* and 0.1 percent fat milk for *L. lactis*.Tryptic soy and Elliker media
were used in liquid, semisolid (0.4% agar) and solid (1.5% agar) forms.

Resuscitation factors: In experiments with *L. lactis* we used the following
factors: mixture of 7 amino acids (glutamine—0.39 g/l, methionine—0.12
g/l, leucine—0.47 g/l, histidine—0.1 g/l, arginine—0.12 g/l, valine—.33 g/l,
isoleucine—0.21 g/l) (Juillard V. et al., 1995); concentratedinactivated biomass
of homologous strain (0.1, 0.5 and 1%); inactivated culture of homologous
strain (5 %) [6]; oleic acid (0.1%, 0.5%, 1% and 5%). Blood substitute
"Aminopeptidum" (0,5 %, 1 % and 10 %) was applied to all species. Media
without factors were used as controls.

22.2.1   VIABILITY AND CULTURABILITYASSAYS

For visual detection of viable and dead bacteria in samples we used Live/Dead
(BacLight™) kit and a luminescence microscope (Karl Zeiss). Culturability was
assessed using plating, most probable number techniques and also by counting
colonies in columns of semisolid agar. For total cell counts we used Goryaev or
Thoma chambers. Number of viable but nonculturable cells were measured as a
result of comparing total viable counts and CFU/ml.

22.3   RESULTS AND DISCUSSION.

For *Lactococcuslactis* resuscitating effect was observed in the following cases (Table 22.1): for strain MSU (after 3.5 months of incubation) addition 1 percent of inactivated, concentrated biomass was to solid Elliker's medium yielded 2.65 (p<0.05) timesincrease in CFU/ml. For the same strain (after 4.5 months of incubation) 3.75 (p<0.05) times increase was noted when 0.5% of biomass was added to liquid Elliker's medium.For strain F-116 we managed to increase culturabilityfour times (p<0.05) by adding mixture of amino acids to semisolid Elliker's medium. Other resuscitating factors (additions to culture media) had no significant effect on nonculturable cells of*L. lactis*. "Aminopeptidum" showed reactivating effect on lyophilized probiotic preparations. For *E. coli* addition of 10 percent of "Aminopeptidum" yielded 6.45 times (p<0.05) increase in culturable counts. For *Lactobacillus acidophilus*culturability increased five times (p<0.05) in the medium with 10 percent of "Aminopeptidum" For *B. bifidum*(see Figure) when 1 percent of"Aminopeptidum" was added value of CFU/ml increased two times (p<0,05) and with addition of 10 percent of "Aminopeptidum" 2.56 times (p<0,05) increase in CFU/ml was observed. It should be noted that addition of "Aminopeptidum" markedlyincreased growth rate of *Lactobacillus acidophilus*. In the medium with the additive colonies formed within 48 hrs while in controls 2/3 of colonies were appeared by 72 hrs (Figure 22.1).

FIGURE 22.1   Stimulating effect of "Aminopeptidum" on *B. bifidum* from bifidumbacterin batch 735. Left – control, middle—1 percent of "Aminopeptidum" right—10 percent of "Aminopeptidum."

TABLE 22.1 Screening different factors for resuscitation of nonculturable cells.

Microorganism	Age of the Culture	Resuscitating Factor	Quantitative characteristics		
			control, KOE/мл	media with factors, KOE/мл	Increase in CFU/ml (times)/p value
L. lactis MSU	3,5мес.	0,1% inactivated, concentrated homologous biomass in Elliker's broth	$0,85 < 4 < 18,7 \times 10^5$	$0,32 < 1,5 < 7,02 \times 10^6$	>0,05
		1% inactivated, concentrated homologous biomass on solid Elliker's medium	$3,65 \pm 0,4 \times 10^5$	$9,68 \pm 1,06 \times 10^5$	**2,65** < 0,05
	4,5мес.	10% "Aminopeptidum" in Elliker's broth	$1,05 < 4 < 15,2 \times 10^5$	$1,84 < 7 < 26,6 \times 10^5$	>0,05
		0,5% inactivated, concentrated homologous biomass in Elliker's broth	$1,05 < 4 < 15.2 \times 10^5$	$0,39 < 1,5 < 5,7 \times 10^6$	**3,75** < 0,05
L. lactis 729			$0,37 < 1,4 < 5,32 \times 10^5$	$0,68 < 2,6 < 9,88 \times 10^5$	>0,05
L. lactis F-116	8 мес.	7 amino acids in semisolid Elliker's broth	$1 \pm 0,11 \times 10^4$	$4 \pm 0,44 \times 10^4$	**4** < 0,05
		10% "Aminopeptidum" in Elliker's broth	1×10^4	6×10^3	>0,05
		5% inactivated homologous culture in Elliker's broth	1×10^4	2×10^3	>0,05
		7 amino acids in 0,1% fat milk	10^3	10^4	>0,05
		10% "Aminopeptidum" in 0,1% fat milk	10^3	10^3	>0,05
		5% inactivated homologous culture in 0,1% fat milk	10^3	10^3	>0,05

TABLE 22.1 (Continued)

	"Aminopeptidum"		10 % of "Aminopeptidum"		
E. coli M-17 (Colibacterin, batch 40–3)			$6,2 \times 10^7$	4×10^8	6,45 < 0,05
Lactobacillus acidophilus (Acipolbatch 11)			$2,2 \pm 2,42 \times 10^5$	$1,1 \pm 1,21 \times 10^6$	5 < 0,05
Bifidobacteri-umbifidum Bifidumbacte-rinbatch 735	1%	9×10^5	$1,81 \times 10^6$	2 < 0,05
	10%	9×10^5	$2,3 \times 10^6$	2,56 < 0,05

It is evident that all probiotic bacteria reacted on addition to "Aminopeptidum." This preparation is a blood substitute so its chemistry is close to the blood. Since probiotics are brought into human or animal intestine and adhere to its walls, they interact with body fluids and thus react to the additive. So we suggest that even if probiotics may contain significant portions of nonculturable cells (particularly ones containing lactobacilli), their effect may be increased when these cells resuscitate after ingestion. For many bacterial species resuscitation via passage through host organism has already been shown ([2].

Role of nonculturable bacteria is not yet sufficiently studied in many areas of microbiology.Researchon this problem may significantly extend understanding of behavior of sanitary significant microorganisms, causes and manifestations of dysbioticconditions,help better characterize pathogenic agents and improve viability assessment of probioticsused for correction of dysbiosises.

22.4   CONCLUSIONS

We conducted a search for factors that promote resuscitation from nonculturable back into actively growing state. For bifidobacteria, *E. coli* and lactobacilli the most efficient was "Aminopeptidum."

KEYWORDS

- *Lactococcuslactis*
- Nonculturable cells
- Probiotics
- Resuscitation

REFERENCES

1. Blinkova, L.; Martirosyan, D.; Pakhomov, Y.; Dmitrieva, O.; Avaughan, R., and Altshuler, M.; *Funct. Food. Health. Dis.* **2014**, *4*(2), 66.
2. Fakruddin, M. D.; Binmannan, K. S.; and Andrews, S.; ISRN Microbiology, **2013**. Article ID 703813, 6 p.
3. Ganesan, B.; Stuart, M. R.; and Weimer, B. C.; *Appl. Environ. Microb.* **2007**, *73*(8), 2498.
4. Hoefman, S., Van Hoorde, K.; Boon, N.; Vandamme, P.; De Vos P.; and Heylen, K.; *PLoS ONE.* **2012**, *7*(4), e34196. doi:10.1371/journal.pone.0034196.
5. Juillard, V.; Le Bars D.; Kunji, E. R. S.; Konings, W. N.; Gripon, J.-C.; Richard, J.; *Appl. Environ Microb.* **1995**, *61*(8) 3024.
6. Miura, M.; Seto Y.; Watanabe M.; and Yoshioka, T.; US Patent Application Publcation, № US2009/0317892 A1/; **2009**.

7. Oliver J. D.; Hite, F.; Mcdougald, D.; Andon, N. L.; and Simpson L. M.; *Appl. Environ. Microb.* **1995,** *61*(7), 2624.

8. Pakhomov, YU. D.; Blinkova, L. P.; Dmitrieva, O. V.; Berdyugina, O. S.; and Stoyanova, L. G.; *J Bacteriol. Parasitol.* **2013,** *5*(1), doi: 10.4172/2155-9597.1000178.

9. Peneau, S.; Chassaing, D.; and Carpentier, B.; *Appl. Environ. Microb.* **2007,** *73*(7), 2839–2846.

10. Senoh, M.; Ghosh-Banerjee, J.; and Ramamurthy, T.; Colwell, R. R.; and Miyoshi, S.; Nair, G. B.; Takeda Y.; *Microbiol. Immunol.* **2012,** *56*(5), 342.

11. Senoh, M.; Ghosh-Banerjee, J.; Ramamurthy, T.; Hamabata T.; Kurakawa, T.; Takeda, M.; Colwell, R. R.; Nair, G. B.; and Takeda, Y.; *Microbiol. Immunol.* **2010,** *54*(9), 502.

12. Steinert, M.; Emody, L.; Amann, R.; and Hacker, J.; *Appl. Environ. Microb.* **1997,** *63*(5) 2047.

CHAPTER 23

IMMUNOLOGICAL DATABASES AND ITS ROLE IN IMMUNOLOGICAL RESEARCH

ANAMIKA SINGH[1] and RAJEEV SINGH[2*]

[1]Maitreyi Collage, University of Delhi,

[2]Division of Reproductive and Child Health Indian Council of Medical Research, New Delhi

*E-mail: 10rsingh@gmail.com

CONTENTS

23.1 INTRODUCTION

Immunological databases are growing day by day as the information related to disease is expanding tremendously. Nowadays Computational immunology expands itself and it is focused on analyzing large scale experimental data and comparison [1, 2]. Immunology related databases cover all other aspects of immune system processes and diseases and the web address which are helpful for epitope designing and new drug designing [3].

Need of Databases development:
1. For the extraction of the existing information of diseases and immune-related resources.
2. Experiment designing on the basis of existing data.
3. Analysis of experiments.
4. Acceleration of knowledge based discovery.

23.1.1 DATABASE DEVELOPMENT

1. The most important aspect of immunology is immunological proteins which are large and they make difference by a single amino acid change and due to this single amino acid change there will be a significant change in the function of the proteins. Due to the databases it is easy to generate a comparative graph between two or more protein with similarity and differences.
2. To understand the origin, structure and function of antibodies MI-IC and other related immunological molecules.
3. To understand the mechanism behind immune disorders, infectious disease, autoimmunity, or tumor immunology.
4. Development and designing of new vaccines and antibodies.

At present large number of databases are available for applied and basic research in immunology. The databases are basically divided into two parts
   (i) **Sequence databases:** Collects the information of protein and DNA, RNA etc [4, 5].
   (ii) **Immunological databases:** Contains information of immune system related proteins and targets [6, 7].

23.1.2 DESCRIPTION OF DATABASES

A database is an organized collection of data. The data are typically organized to model relevant aspects of reality in a way that supports processes requiring this information. These general sequence databases are essential for molecular immunology projects because they provide interesting hits and useful insights about a particular sequence of immunological interest. For getting more information, there are many other specialized immunological databases. This lecture provides brief description about each immunological database.

Antigen DB (http://www.imtech.res.in/raghava/antigendb/): Sequence, structure, and other data on pathogen antigens [8].

IMGT (http://www.ebi.ac.uk/imgt): It contains two databases, IMGT/LIGM-DB, a comprehensive database of Ig and TeR sequences from human and other vertebrates, and IMGT/HLA-DB, a database of human MHC. It enables users to extract data on nucleotide and protein sequences, sequence alignment, alleles, sequence tagged sites and polymorphisms, gene maps and genetic data, structural data, oligonueleotide primers, relationship with disease and cell lines [9].

FIMM (http://sdmc.krdl.org.sg.8080/gimm) : focuses on cellular immunology, specifically on MHC, antigenic rpoteins, antigenic peptides, and relevant disease information. The tools include keyword search, pattern search, BLAST searches, multiple sequence alignment, and binding pocket/contact sites analysis [10].

VBASE (http://www.mrc-cpe.cam.ac.uk/imt-doc/public): contains germ line variable region sequences of human antibodies. The search tool at this site helps to obtain amino acid and nucleotide sequences, scale maps of the human immunoglobulin loci, sequence alignments, numbers of functional segments, restriction enzyme cuts in V genes, and PCR primers for rearranged V genes [11].

Hybridoma Data Bank (HDB) (http://www.atcc.org/hdb/hdb.litrnl): This database contains information on hybridomas and other cloned cell lines and their immunoreactive products (e.g. monoclonal antibodies). A HDB record contains comprehensive information on cloned cell lines including bibliography, biological origin, classification, methodological description, reactivity details, distributors, applications, availability and other relevant details [12].

MHCPEP (http://bio.dfci.harvard.edu/DFRMLI/): This database contains list of MHC-binding peptides [13].

BCIPEP (http://www.imtech.res.in/raghava/bcipep): The BCIPEP is a database of immunodominant peptides, which result in stimulation of B-cell lineage. The database is consisting of nearly ~1100 B-cell epitopes collected from literature. The data is kept in following fields: peptide sequence, antibody used for testing, reference, database reference of parent antigenic protein, Measure of antigencity and immunogenicity [14].

SYFPEITHI (http://www.uni-tuebingen.de/uni/kxi): It is a database of MHC ligands and peptide motifs. Users can extract individual binding motifs and related peptides or search the database by peptide sequence. Additional options include search by anchor positions, peptide source, or peptide mass [15].

MHCDB   (http://www.limp.mrc.ac.uk/Registered/Option/mhedb.Html): The database contain physical and genetic maps of human major histocompatibility complex that include fully annotated genomic DNA sequences, cDNA sequences of class I and class II alleles [16].

HPTAA (http://www.bioinfo.org.cn/hptaa/): HPTAA is a database of potential tumor-associated antigens that uses expression data from various expression platforms, including carefully chosen publicly available microarray expression data, GEO SAGE data and Unigene expression data. [17].

HIV Molecular Immunology Database (http://hiv-web.lanl.gov/immuno-lov/index.html): It contains an annotated, searchable collection of HIV-1 cytotoxic and helper T-cell epitopes and antibody binding sites. The search tools include motif/pattern searches, sequence alignment to all sequences in the HIV-I genome and BLAST searches. The main aim of database is to provide comprehensive listing of HIV epitopes [18].

Epitome (https://rostlab.org/services/epitome/) is a database of all known antigenic residues and the antibodies that interact with them, including a detailed description of the residues involved in the interaction and their sequence/structure environments. Each entry in the database describes one interaction between a residue on an antigenic protein and a residue on an antibody chain. Every interaction is described using the following parameters: PDB identifier, antigen chain ID PDB position of the antigenic residue, type of antigenic residue and its sequence environment, antigen residue secondary structure state, antigen residue solvent accessibility, antibody chain ID, type of antibody chain (heavy or light), CDR number, PDB position of the antibody residue, and type of antibody residue and its sequence environment. Additionally, interactions can be visualized using an interface to Jmol [19].

23.1.3   INTERFERON STIMULATED GENE DATABASE

(http://www.lerner.ccf.org/labs/williams/xchip-html.cgi): Interferons (IFN) are a family of multifunctional cytokines that activate transcription of a subset of genes. The gene products induced by IFN are responsible for the antiviral, antiproliferative and immunomodulatory properties of this cytokine. The database is fully searchable and contains links to sequence and Unigene information. The database and the array data are accessible via the World Wide Web.

MHCBN: The data of experimentally proven MHC binders, MHC nonbinders and T-cell epitopes is a prime requirement for the development of prediction method for T-cell epitope and/or MHC binders. The achievement of this goal is possible through the development of a comprehensive database in cellular immunology [20].

In the past, a number of databases have been created to provide information about MHC-binding peptides and T-cell epitopes. The databases like SY-FPEITHI, JenPep, and HIV Database are modest in size and provide very focused information. Another database, FIMM contains information about MHC associated peptides, antigens, MHC molecules and associated disease. However FIMM provides a rich set of internal/external data links and extraction of complex information, but it maintains only about 1,500 naturally processed peptides or T-cell epitopes. The MHCPEP is a widely used database that contains information about 13400 MHC-binding peptides. It has greater proportional coverage than any of above mentioned databases. The main limitations of MHCPEP are i) it has not been updated since1998 ii) database has no tools for data extraction/analysis and iii) it is not linked with other database.

MHCBN Tools for extraction and analysis of data

The database has a set of web tools for extraction and analysis of data. The data extraction tools include **general query tool** and **peptide search tools** for making complex queries. The tools for the analysis of the data include

i)   Tools for creation of datasets.

ii)   MHC BLAST

iii) Antigenic BLAST

iv) Mapping of antigenic regions in query sequence

v)   Online submission of data.

The tool for mapping of antigenic regions is very useful tool for locating promiscuous antigenic regions in the query sequence.

23.1.4   BASED ON IMMUNOLOGICAL DATABASES

Reference Database of Immune Cells (RefDIC): RefDIC is an open resource of quantitative mRNA/Protein profile data specifically for immune cells [21, 22].

Innate Immune Database (IIDB): IIDB is a repository of computationally predicted transcription factor binding sites for over 2000 mouse genes associated with immune response behavior. A specific focus of IIDB is on Toll-like Receptor (TLR) genes, which are key components of innate immunity.

Immunology Database and Analysis Portal (ImmPort): The ImmPort system provides information technology support in the production, analysis, archiving, and exchange of scientific data for researchers supported by NIAID/DAIT. It serves as a long-term, sustainable archive of data generated by investigators funded through the NIAID/DAIT. The ImmPort system also provides data analysis tools and an immunology-focused ontology.

Case studies based on Immunology Database and Analysis Portal (ImmPort):

Study Title	PI	Type of Ex...	Public Rel..
⊟ Atopic Dermatitis & Vaccinia Network (ADVN) (14 Studies)			
SDY6: ADVN Biomarker Registry Study	Lisa Beck	ELISA	11/16/2012
SDY8: ADVN Biomarker Registry Study: CMI-HSV Substudy	Donald Leung	ELISPOT,EL...	11/16/2012
SDY7: ADVN Biomarker Registry Study: CMVAb-Vaccinia Substudy	Donald Leung	-	11/16/2012
SDY9: ADVN Biomarker Registry: Neutrophil Substudy	Lisa Beck	FCM,ELISA	11/16/2012
SDY5: Analysis and Correlation of Cathelicidin Expression in Skin and Saliva...	Richard Gallo	-	11/16/2012
SDY13: Analysis of the Response of Subjects with Atopic Dermatitis to Oral...	Donald Leu...	-	11/16/2012
SDY14: Antimicrobial Response to Oral Vitamin D3 in Patients with Psoriasis	Richard Gallo	-	11/16/2012
SDY4: Risk Factors in Atopic Dermatitis for the Development of Eczema Her...	Thomas Bie...	FCM	11/16/2012
SDY10: Role of Antimicrobial Peptides in Host Defense Against Vaccinia Virus	Donald Leung	-	11/16/2012
SDY11: Genetics of Atopic Dermatitis-Eczema Herpeticum	Lisa Beck, ...	-	
SDY12: Pilot Study to Determine the Underlying Mechanisms for Infection an...	Donald Leung	-	
SDY2: Immune Response to Varicella Vaccination in Subjects with Atopic D...	Lynda Sch...	FCM,ELISP...	11/16/2012
SDY3: Responses to Immunization with Keyhole Limpet Hemocyanin (KLH) ...	Henry Milgrom	-	11/16/2012

Immunogenetic Related Information Source (IRIS): IRIS is a database of all known human defense genes, produced in the laboratory of Professor John Trowsdale at the University of Cambridge. IRIS currently includes chromosomal locations, functional annotations, and sequence data for over 1,500 functional human immune genes. Please note that the IRIS database does not seem to be available anymore.

Immunome Database for Genes and Proteins of the Human Immune System: Immunome contains information about immune-related proteins, their domain structure and related ontology terms. Information can also be found for the localization of the coding genes and their comparison with the existing mouse orthologs [23, 24].

Macrophages.com: Macrophages.com is an online resource for those interested in macrophages and their role as major effector cells in innate and adaptive immunity. This website is designed to act as a centralized resource for the worldwide community of scientists interested in different aspects of macrophage biology [25].

23.1.5   PROJECTS BASED IMMUNOLOGICAL DATABASES

DC ATLAS project: DC ATLAS is an immunological and bioinformatics integrated project, developed as a joint effort within the DC-THERA European Network of Excellence (www.dc-thera.org), a collaborating network established under the European Commission's Sixth Framework, programmed to translate discoveries from DC immunobiology into clinical therapies. The major scientific and technological goal of DC ATLAS is to generate complete maps of the intracellular signaling pathways and regulatory networks that govern DC maturation/activation and function.

Immunological Genome Project (ImmGen): The Immunological Genome Project is a collaborative group of Immunologists and Computational Biologists who are generating, under carefully standardized conditions, a complete microarray dissection of gene expression and its regulation in the immune system of the mouse. The project encompasses the innate and adaptive immune systems, surveying all cell types of the myeloid and lymphoid lineages with a focus on primary cells directly ex vivo.

23.2 CONCLUSION

These databases provide information for understanding the specificity of immune system and immunological response at molecular level and designing a rapid in silico vaccine. Some of these databases provide special tools for in silico vaccine designing. The tools for detecting immunodominant region in an antigenic sequence are available at some databases e.g. antigenic mapping at MHCBN. The tools for identifying an immunological protein are available as BLAST against MHC database or antigenic protein database. Some database like MHCPEP, MHCBN, FIMM, SYFPEITHI provide a collection of MHC-binding peptides or T-cell epitopes which form the basis of prediction methods for subunit vaccine design. Databases like SYFPEITHI, FIMM has tools for epitope prediction which are used for novel epitope prediction. The databases which provide structural information about TCR, MHC, and antigenic sequences are useful for making structure-based prediction methods and rational drug designing. The blast tools available at various databases can be used to detect conserved epitopic region in different strains of pathogens.

KEYWORDS

- **Description of databases**
- **Immunological database**
- **Sequence database**

REFERENCES

1. Tong, J. C.; and Ren, E. C.; Immunoinformatics: current trends and future directions. *Drug. Discov*. **2009**, *14* (13–14), 684–689.
2. Korber, B.; LaBute, M.; and Yusim, K.; Immunoinformatics comes of age. *PLoS. Comput. Biol*. 2(6), e71.
3. Ross, R.; (1 February 1916); An application of the theory of probabilities to the study of a priori pathometry. Part I" (PDF). *Proc. Royal. Soc. Lond. Ser*. A *92*(*638*), 204–230.
4. Sikic, K.; and Carugo, O. Protein sequence redundancy reduction: comparison of various method. *Bioinformation*. 5(6), 234–239. PMID 21364823.
5. Iliopoulos, I.; Tsoka, S.; Andrade, M. A.; Enright, A. J.; Carroll, M.; Poullet, P.; Promponas, V.; and Liakopoulos, T. et al. Evaluation of annotation strategies using an entire genome sequence. Bioinformatics. **2003**, *19*(6), 717–726. PMID 12691983
6. Tong, J. C.; and Ren, E. C.; (July 2009). "Immunoinformatics: current trends and future directions". *Drug Discov. Today*. *14*(13–14), 684–689. doi:10.1016/j.drudis.2009.04.001. PMID 19379830.
7. Korber, B.; LaBute, M.; and Yusim, K.;" Immunoinformatics comes of age". *PLoS Comput. Biol*. **2006**, *2*(6), e71.
8. Ansari, H. R.; Flower, D. R.; Raghava, G. P.; AntigenDB: an immunoinformatics database of pathogen antigens. Nucleic Acids Res. **2010**, *38*, (Database issue), D847–53. doi:10.1093/nar/gkp830.PMC 2808902. PMID 19820110.
9. Lefranc, M. P.; (January 2001). "IMGT, the international ImMunoGeneTics database". *Nucl. Acid. Res*. *29*(1), 207–209. PMC 29797. PMID 11125093.
10. Schönbach, C.; Koh, J. L.; Flower. D. R.; and Brusic, V.; An update on the functional molecular immunology (FIMM) database. *Appl Bioinformatics*. **2005**, *4*(1), 25–31.
11. Retter, I.; Althaus, H. H.; Münch, R.; Müller, W.; (January 2005). "VBASE2, an integrative V gene database". *Nucl. Acid. Res*. *33*(Database issue), D671–D674.
12. Nihon Rinsho.; **1992**, *50*(11), 2808–2815.
13. Zhang, G. L.; Lin, H. H.; Keskin, D. B.; Reinherz, E. L.; and Brusic, V.; Dana-Farber repository for machine learning in immunology. *J. Immunol. Methods*. **2011**, *374*(1–2), 18–25.
14. Saha, S.; Bhasin, M.; and Raghava, G. P.; Bcipep: a database of B-cell epitopes. *BMC Genom*. **2005**, *6*, 79. doi:10.1186/1471-2164-6-79.
15. Rammensee, H.; Bachmann, J.; Emmerich, N. P.; Bachor, O. A.; and Stevanović, S.; SYFPEITHI: database for MHC ligands and peptide motifs. *Immunogenetics*. **1999**, *50*(3–4), 213–219.
16. Greenbaum, J. A.; Andersen, P. H.; Blythe, M.; Bui, H. H.; Cachau, R. E.; Crowe, J.; Davies, M.; Kolaskar, A. S.; Lund, O.; Morrison, S.; Mumey, B.; Ofran, Y.; Pellequer, J. L.; Pinilla, C.; Ponomarenko . J. V.; Raghava, G. P.; Regenmortel, M. H.; Roggen, E. L.; Sette, A.; Schlessinger, A.; Sollner, J.; Z and M, Peters, B.; Towards a consensus on datasets and evaluation metrics for developing B-cell epitope prediction tools. *J. Mol. Recognit*. **2007**, *20*(2), 75–82.
17. Wang, X.; Zhao, H.; and Xu, Q.; et al. HPtaa database-potential target genes for clinical diagnosis and immunotherapy of human carcinoma. *Nucl. Acid. Res*. **2006**, *34* (Database issue), D607–D612.
18. Korber, B. T. M.; Brander, C.; Haynes, B. F.; Koup, R.; Moore, J. P.; Walker, B. D.; Watkins, D. I.; HIV Molecular Immunology 2006/2007. Los Alamos, New Mexico: Los Alamos National Laboratory, Theoretical Biology and Biophysics; **2007**.
19. Schlessinger, A.; Ofran, Y.; Yachdav, G.; and Rost, B.; Epitome: database of structure-inferred antigenic epitopes. *Nucl. Acid. Res*. **2006**, *34*(Database issue), D777–D780.
20. Bhasin, M.; Singh, H.; and Raghava, G. P.; MHCBN: a comprehensive database of MHC binding and non-binding peptides. Bioinformatics. **2003**, *19* (5), 665–656.

21. Hijikata, A.; Kitamura, H.; Kimura, Y.; Yokoyama, R.; Aiba, Y.; Bao, Y.; Fujita, S.; Hase, K.; Hori, S.; Kanagawa, O.; Kawamoto, H.; Kawano, K.; Koseki, H.; Kubo, M.; Kurita-Miki, A.; Kurosaki, T.; Masuda, K.; Nakata, M.; Oboki, K.; Ohno, H.; Okamoto, M.; Okayama, Y.; O-Wang, J.; Saito, H.; Saito, T.; Sakuma, M.; Sato, K.; Seino, K.; Setoguchi, R.; Tamura, Y.; Tanaka, M.; Taniguchi, M.; Taniuchi, I.; Teng A.; Watanabe, T.; Watarai, H.; Yamasaki, S.; and Ohara, O.; Construction of an open-access database that integrates cross-referenceinformation from the transcriptome and proteome of immune cells. *Bioinformatics.* **2007**, *23*, 2934–2941.
22. Kimura, Y.; Yokoyama, R.; Ishizu, Y.; Nishigaki, T.; Murahashi, Y.; Hijikata, A.; Kitamura, H.; and Ohara, O.; Construction of quantitative proteome reference maps of mouse spleen and lymph node based on two-dimensional gel electrophoresis. *Proteomics.* **2006,** *6*, 3833–3844.
23. Ortutay, C.; Siermala, M.; and Vihinen, M.; Molecular characterization of the immune system: emergence of proteins, processes, and domains. *Immunogenetics.* doi:10.1007/s00251-007-0191-0
24. Ortutay, C.; and Vihinen, M.; Immunome: A reference set of genes and proteins for systems biology of the human immune system. *Cell Immunol.* 2007.
25. Mossadegh-Keller, N.; Sarrazin, S.; Kandalla, P. K.; Espinosa, L.; Stanley, E. R.; Nutt, S. L.; Moore, J.; and Sieweke, M. H.; M-CSF instructs myeloid lineage fate in single haematopoietic stem cells. *Nature.* **2013**, *497*(7448), 239–243.

CHAPTER 24

EFFECT OF ALKYLRESORCINOLS ON THE CHITOLYTIC ACTIVITY OF LYSOZYME AND PAPAIN

E. I. MARTIROSOVA, N. A. GREBENKINA, and I. G. PLASHCHINA

[1]Emanuel Institute of Biochemical Physics, RAS Kosygina st., 4 Moscow, 119334, E-mail: ms_martins@mail.ru

[2]Higher Chemical College, RAS, Miusskaya sq., 9 Moscow, 125047

CONTENTS

24.1 INTRODUCTION

Chitosan is a *polyaminosaccharide,* a partially deacetylated **derivative** of chitin. It is widely used in food, biomedical and chemical industries, as fat blockers, a stabilizer, a preservative for fruits and vegetables, dairy products. From the chemical point of view the structure of chitosan is a copolymer of glucosamine and N-acetylglucosamine linked by β-1,4-glycosidic bonds. Despite the high biofunctional properties of chitosan, its use is limited because of the high molecular weight and viscosity and, as a result, low absorption *in vivo.* Products of chitosan depolymerization, low molecular weight derivatives as well as chitooligomers exhibit physiological activity greater than chitosan, and therefore have a great potential of application. Low molecular weight chitosans (LMWC) with molecular weight between 5 and 10 kDa shows a strong antibacterial, fungicidal, hypolipidemic, and hypocholesterolemic effect [1; Kumar et al., 2004; 3].

Sphere of chitosan and its low molecular weight derivatives application is constantly expanding. They are used in agriculture as a component of livestock feed, which increases resistance to disease, as part of fertilizers, as a means of prebactericidal seed processing. LMWC can be obtained by physical, chemical or biological methods. Last method is preferred because it more specific and can be easy controlled by regulation of pH, reaction time and temperature. Biological methods of chitosan destruction can be divided conditionally into three groups: 1. with using of micromushrooms destructors; 2. with using of purified specific chitolytic enzymes of microorganisms; 3. with using of another hydrolase classes, nonspecific to given substrate—lipases, glycosidases (lysozyme), and amidases. In the first and second cases, microbial chitinases are used routinely from cultural media directly or separated enzyme preparates. As it was shown, some nonspecific hydrolases are capable to chitosan depolymerization with the same efficiency as chitinases. At the same time a range of amidhydrolases (pepsin, papain, bromeline, and ficin) can depolymerize chitosan even more effectively than chitinases [4].

Often enzymatic hydrolysis fails to provide a high degree of substratum hydrolysis because of the enzyme inactivation. Rise of the enzymatic activity and functional stability is a crucial problem of current biotechnology. One of the most effective ways of modifying enzymes is utilizing their weak nonspecific interactions with low molecular ligands. Microbial low molecular extracellular metabolites performing the functions of autoinducers of anabiosis represent one of the groups of biologically active substances capable of affecting the activity and stability of lysozyme. These autoregulators represented in a number of bacteria and yeasts by alkylhydroxybenzenes (AHB), alkylresorcinols in particular, induce transition of microbial cells into a hypometabolic (anabiotic) state and realize this function through interaction with a wide variety of biopolymers of a bacterial cell [5, 6]. Nonspecific influence of these autoregulators on enzymatic proteins is associated with the chemical structure of AHB and the type of their interplay with protein molecules [12].

As was shown earlier, modification of some hydrolytic enzymes by chemical analogs of microbial autoregulatory factors is capable of raising the activity of enzymes *in vitro*, increasing the depth of hydrolysis of the industrial substrate, and also expanding the temperature and pH-ranges of catalysis [7].

First part of our work is devoted to researching the chitolytic activity of lysozyme for chitosan hydrolysis. Lysozyme is widely used in medicine and food industries as antimicrobial agent. During last 20 years lysozyme is intensively used for production baby food and nutrition.

It was shown earlier that efficiency of lysozyme hydrolyses of different substrates can be increased due to methylresorcinol, which is the simplest representative of alkylresorcinols. 5-Methylresorcinol has been noted to stimulate the lysozyme activity within the range of concentrations $10^{-7} - 10^{-3}$ M up to 120 percent when peptidoglycane from the *Micrococcus luteus* is used as a substrate. When nonspecific substrates (colloid chitin, *Saccharomyces cerevisiae* cells) are used, then the growth of hydrolytic activity was 470 percent and 400 percent, correspondingly [8]. So, MR shows the ability to change a substrate specificity which is appeared in increasing of hydrolysis rate of the bounds nonspecifically for this enzyme [8]. In our work the effect of MR on the lysozyme chitolytic activity is shown with using of homogeneous substrate chitosan.

Another part of the work is to study the effect of AR on the chitolytic activity of papain. Papain is a potent proteolytic enzyme of plant origin, belonging to the family of cysteine proteases. Its enzymatic and physiological properties are the subject of series studies, because it plays an important role in the physiology of plants and it is widely used in the food and pharmaceutical industries. In particular, the papain used in the food industry for the softening of meat, production of hydrolysates, clarification of juices and beer, extraction of color and odor plants components, in the dairy industry for cheese manufacture.

24.2   THE AIM OF THIS WORK

The aim of this work was to study the effects of alkylresorcinols (methylresorcinol and **hexylresorcinol) on the** enzymatic activity of two types of hydrolases such as glycosidases (lysozyme) and amidhydrolases (papain) for chitosan hydrolysis.

24.3   MATERIALS AND METHODS

A sample of hen egg white lysozyme (Sigma-Aldrich, USA) with molecular mass 14 445 g/mol, papain from Papaya carica (Merk, USA) with molecular mass 23 500 g/mol were used. Alkyl-substituted hydroxybenzenes, 5-methyl-resorcinol monohydrate (5-Methylbenzene-1,3-diol) (Sigma-Aldrich, USA) with molecular mass 142 g/mol and 4-hexylresorcinol (Sigma, USA) with molecular mass 194 g/mol were taken. Substrate for both enzymes was chitosan with deacetylation degree 58±3 percent. It was prepared from commercial chitosan (Sigma-Aldrich) with deacetylation degree 85 percent.

24.3.1   SOLUTIONPREPARATION

Protein powder was dissolved in buffer for 2 hrs and then centrifuged at 20 000g (Beckman 21, Germany) during 1 hr at room temperature. The operating solutions were prepared by mixing equal volumes of protein and alkylresorcinol solutions within 40–50 min just before using in experiment (preincubation time).

Methylresorcinol was dissolved in pure 0.05 M acetic buffer. Hexylresorcinol was dissolved in ethanol solution and then 0.05 M acetic buffer was added to final alcohol concentration 10%. In the control systems, an equivalent amount of the solvent was used instead of the AR solution.

24.3.2   DETERMINATION OF ENZYMATIC ACTIVITY

Chitosan solution (1%; dissolved in 0.05M acetic buffer, pH 4.5 for lysozyme and 5.0 for papain) was separately treated with modified by AR enzymes (i.e. papain and lysozyme) in the ratio 15:1 (w/w) for lysozyme and 25:1 (w/w) for papain, incubated for 3–4 h at temperatures 37–45°C, followed by arresting the reaction by heat-denaturing the enzyme (100°C, 10 min) and adjusting the pH to 10.0 using 2 M NaOH. The supernatant was separated by centrifugation (10,000 g and 15 min). The concentration of acetylglucosamine, the product of chitosan hydrolysis, was determined in supernatant by the procedure with dinitrosalicylic acid (DNSA) [9].

24.4   RESULT AND DISCUSSION

24.4.1   EFFECT OF ALKYLRESORCINOLS ON THE LYSOZYME ACTIVITY

It was established that the use as methylresorcinol (MR), and hexylresorcinol (HR) let to increase the lysozyme activity in the whole concentration range of the used modifiers.

The curve of lysozyme activity (% of control) as a function of the MR concentration has a bimodal shape (Figure 24.1), which is typical for many hydrolases [7]. The first peak corresponds to a MR concentration 0.1 mg/ml and was 60 percent. The second part of the curve has the shape of a curve with saturation.

Maximal effect of lysozyme activity enhancing was attained at a MR concentration 2.0 mg/ml (100%). It can be assumed that the presence of the second peak of lysozyme activity in the chitosan hydrolysis reaction is of due to the interaction of lysozyme with ligand molecules in partially associated form. Earlier using methods of isothermal microcalorimetry and dynamic tensiometry it was established the fact of MR self-organization in solution. The value of micelle formation concentration or self-organization (conditionally, CMC) at pH 6.0 is 2.36 mg/ml [10, 11]. Furthermore, the synergistic effect of interaction the MR with lysozyme in solution, resulting in displacement of CMC_{MR} in the protein presence was found using a method of dynamic tensiometry. Clarification of CMC_{MR} values depending of pH in the absence and presence of lysozyme is currently produced.

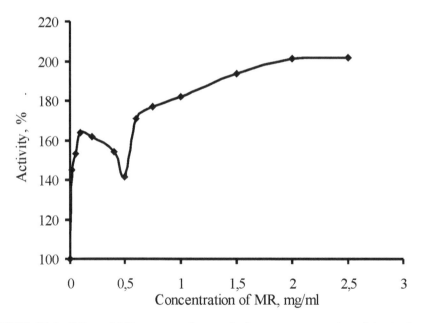

FIGURE 24.1 Effect of MR concentration on the lysozyme enzymatic activity in the process of chitosan hydrolysis. Experiment parameters: 37°C, pH 4.5, 3 hrs, E/S = 1/15.

In the case of HR, which has more alkyl chain length, hence the higher hydrophobicity, the dependence of the activity from HR concentration has a form of a curve with saturation (Figure 24.2). Apparently, the effect is due, as in the

MR case, the transition of the molecular form to HR micelle one which restricts access of modifier into the active site of the enzyme. In this case, the maximum effect was 55 percent of control at a HR concentration of 1.12 mg/ml.

Earlier it was shown that HR has a surface activity. In the 5 percent ethanol solution its CMC is 0.9 mg/ml [5, 6]. In this experiment, saturation region was at HR concentrations 0.7–1.0 mg/ml (in preincubation stage), apparently due to the presence not only the molecular but micelle form of HR in the solution. The micelle form can serve as a barrier for penetration of the HR in the enzyme active center, and therefore do not provide additional stimulatory action to the enzyme activity.

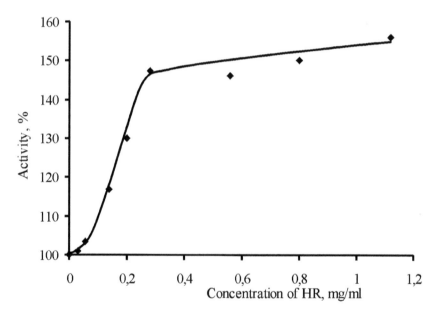

FIGURE 24.2   Effect of HR concentration on the lysozyme enzymatic activity in the process of chitosan hydrolysis. Experiment parameters: 37°C, pH 4.5, 3 hrs, E/S = 1/15.

24.4.3   EFFECT OF ALKYLRESORCINOLS ON THE PAPAIN ACTIVITY

Figure 24.3 shows the dependence of chitolytic papain activity from MR concentration in the hydrolysis reaction of the chitosan.

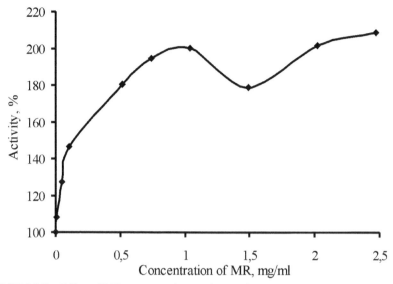

FIGURE 24.3 Effect of MR concentration on the papain enzymatic activity in the process of chitosan hydrolysis. Experiment parameters: 45°C, pH 5.0, 4 hrs, E/S = 1/25.

Activity curve as in the case of lysozyme has a bimodal character. The position of the first maximum corresponds to the concentration of MR 1.0 mg/ml, which is 10 times higher than for lysozyme. The increase of activity in this case was 100 percent. A similar increase the activity value (100%) is observed in the second peak at MR concentration 2.0–2.5 mg/ml.

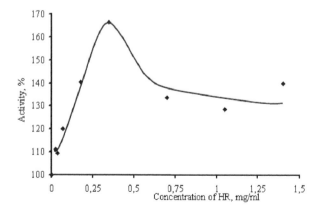

FIGURE 24.4 Effect of HR concentration on the papain enzymatic activity in the process of chitosan hydrolysis. Experiment parameters: 45°C, pH 5.0, 4 hrs, E/S = 1/25.

HR application for papain modification led to increase enzyme activity maximum up to 65% at a ligand concentration 0.3 mg/ml (Figure 24.4). The highest value of the enzymatic activity, as in the case with the lysozyme modification by HR corresponds to the concentration range of 0.7–1.0 mg/ml (in preincubation stage), the limit value of which corresponds to CMC_{HR}. Further increasing of the modifier concentration in the medium led to a slight decrease effect of the enzymatic activity.

24.5 CONCLUSION

The performed experiments provide evidence in favor of the ability of alkylresorcinols—chemical analogs of the homologs of autoinducers of bacterial anabiosis to modify the functional activity of lysozyme and papain and to regulate the efficiency of chitosan hydrolysis. The efficiency of the enzymes modification has been shown to depend on the AR concentration and its structure. The obtained results substantially conform to the data demonstrating the effects of AR on other mono-subunit enzymes (trypsin, α-, β- and glycoamylase, ribonuclease, etc) and confirm the capability of AR to modify enzymatic proteins nonspecifically to their structure [7, 12]. Changes in the activity of the enzymes in their complexes with AR can be the result of the increasing intramolecular dynamic of enzyme-AR complex, reflected in the enhancing of the equilibrium fluctuations amplitudes [13] and in the decreasing of thermodynamic stability of the protein molecule [14].

Shape of enzyme activity dependence curve from AR concentration correlates with AR self-organizing conditions in the solution. This question requires of additional research.

The experiments have demonstrated the ability of AR to enhance the functional activity of hydrolytic enzymes and thus to increase the efficiency of enzyme in the chitosan hydrolysis.

KEYWORDS

- **Alkylresorcinols**
- **Chitosan**
- **Hexylresorcinol**
- **Lysozyme**
- **Methylresorcinol**
- **Papain**

REFERENCES

1. Vishu Kumar, A. B.; and Tharanathan, R. N.; A comparative study on depolymerization of chitosan by proteolytic enzymes. *Carbohydrate. Polym*. **2004**, *58*(3), 275–283.
2. Vishu Kumar A. B.; Varadaraj, M. C.; Gowda, L. R.; and Tharanathan, R. N.; Low molecular weight chitosans—Preparation with the aid of pronase, characterization and their bactericidal activity towards Bacillus cereus and Escherichia coli. *Biochimica et Biophysica Acta*. **2007**, 1770, 495–505.
3. Shih-Bin Lin; Yi-Chun Lin; and Hui-Huang Chen; Low molecular weight chitosan prepared with the aid of cellulase, lysozyme and chitinase: Characterization and antibacterial activity. *Food Chem*.**2009**, *116*, 47–53.
4. Frolov, V. G.; Dyshkova, Z. G.; and Cherkasova, E. I.; Investigation of chytolytic activity of papain. Materials of 8th International conference Modern prospects in chytin and chytosan investigation. Kazan. 2006 June 12–17. pp 315–318 (in Russian)
5. Bespalov, M. M.; Kolpakov, A. I.; Loiko, N. G.; Doroshenko, E. V.; Mulyukin, A. L.; Kozlova, A. N.; Varlamova, E. A.; Kurganovm, B. I.; and El'-Registan, G. I.; The role of microbial dormancy autoinducers in metabolism blockade. *Microbiology*. **2000**, *69*(2), 217–223.
6. Kolpakov, A. I.; Il'inskaya, O. N.; Bespalov, M. M.; Kupriyanova-Ashina F.G., Gal'chenko V.F., Kurganov V.I., El'-Registan G.I. Stabilization of Enzymes by Dormancy Autoinducers as a Possible Mechanism of Resistance of Resting Microbial Forms. Microbiology. **2000**, *69*(2), 180–185.
7. Martirosova, E. I.; Karpekina, T. A.; and El'-Registan, G. I.; Enzyme modification by natural chemical chaperons of microorganisms. *Microbiology*. **2004**, *73*(5), 609–615.
8. Petrovskii, A. S.; Deryabin, D. G.; Loiko, N. G.; Mikhailenko, N. A.; Kobzeva, T. G.; Kanaev, P. A.; Nikolaev, Yu. A.; Krupyanskii, Yu. F.; and Kozlova, A. N.; El'-Registan G.I. Regulation of the functional activity of lysozyme by alkylhydroxybenzenes. *Microbiology*. **2009**, *78*(2), 146–155.
9. Miller, G. L.; Use of dinitrosalicylic acid reagent for determination of reducing sugar. *Anal. Chem*. **1959**, *31*(5), 426–428.
10. Martirosova, E. I.; Regulation of hydrolase catalytic activity by alkylhydroxybenzenes: thermodynamics of C_7-AHB and hen egg white lysozyme interaction. Biotechnology and the Ecology of Big Cities. Nova Science Publishers: New-York, USA, **2011**; 105–113.
11. Martirosova, E. I.; and Plashchina, I. G.; Improvement of the functional properties of lysozyme by interaction with 5-methylresorcinol. Pharmaceutical and medical biotechnology. New perspectives. Nova Science Publishers: New-York, USA, 2013, 45–54.
12. Nikolaev, Yu.A.; Loikom, N. G.; Stepanenko, I.Yu.; Shanenko, E. F.; Martirosova, E. I.; Plakunov, V. K.; Kozlova, A. N.; Borzenkov, I. A.; Korotina, O. A.; Rodin, D. S.; Krupyanskii Yu.F.; El'-Registan, G. I.; Changes in physicochemical properties of proteins, caused by modification with alkylhydroxybenzenes. *Appl. Biochemy. Microbiol*. **2008**, *44*(2), 143–150.
13. Krupyanskii, Y. F.; Abdulnasirov, E. G.; Korotina, O. A.; Stepanov, S. A.; Knox, P. P.; Zakharova, N. I.; Rubin, A. B.; Loiko, N. G.; and Nikolaev, Y. A.' El'-Registan G.I. Influence of chemical chaperones on the properties of lysozyme and the reaction center protein from Rhodobacter sphaeroides. *Biophysics*. **2011**, *56*(1), 8–23.
14. Plashchina, I. G.; Zhuravleva, I. L.; Martirosova, E. I.; Petrovskii, A. S.; Loiko, N. G.; Nikolaev Yu.A.; and El'-Registan, G. I.; Effect of Methylresorcinol on the Catalytic Activity and Thermostability of Hen Egg White Lysozyme. Biotechnology, Biodegradation, Water and Foodstuffs, Nova Science Publishers: New-York, USA, **2009**; 45–57.

CHAPTER 25

A RESEARCH NOTE ON BIOTECHNOLOGICAL PREPARATIONS FOR ENHANCING THE QUALITY OF DOMESTIC FISH MIXED FEED

D. S. PAVLOV[1], N. A. USHAKOVA[1], V. G. PRAVDIN[2],
L. Z. KRAVTSOVA[2], C. A. LIMAN[3], and S. V. PONOMAREV[4]

[1]A.N. Severtsov Institute of Ecology and Evolution, Russian Academy of Sciences, 119071 Russia, Moscow, Leninskij prosp., 33, E-mail naushakova@gmail.com

[2]The "NTC BIO", LLC, 309292 Russia, Belgorod region, Shebekino town, Dokuchayev str., 2, E-mail: ntcbio@mail.ru

[3]The "Agroakademia", LLC, 309290 Russia, Belgorod region, Shebekino town, A. Matrosov str. 2A, E-mail: agroakademia@mail.ru

[4]The "Bioaquapark" Innovation Centre– the Scientific Centre of the Aqua-Culture at the ASTU, 414025, Astrakhan, Tatischev str., 16, E-mail: kafavb@yandex.ru

CONTENTS

In order to improve the quality of fish mixed feed, some enzyme, probiotic, prebiotic and probiotic/enzyme combination feed additives are used, as well as complex probiotic preparations enriched with phytocomponents.

Probiotic fodder preparations are regarded as a potential alternative to feed antibiotics, so the use of probiotics is considered an essential point of *obtaining* ecologically clean feed [1–5]. Probiotic preparations balanced with phytochemicals show an enhanced biological activity due to the combination of the actual probiotic effect and the action of a phytobiotic.

Probiotics are live microbial supplements that have a beneficial effect on the body by improving the intestinal microbial balance, and stimulate the metabolism and immune processes. Probiotics are widely used in mixed feed for fish [6–9]. In themselves, probiotics do not provide a significant amount of nutrients for producing more products. But their biological potential improves fish health, enhances productivity levels, and better use of feed.

The determining factor of the probiotics efficiency is, in many ways, the technology of formulating these preparations. Modern biotechnology approaches to the development of probiotic preparations imply, firstly, the use of different types of microorganisms in certain combinations, and, secondly, their production in a form allowing their long-term storage at normal temperatures, and granulation.

The technology for production of the biologically active complex probiotic preparations ProStor and Ferm KM-1 is based on a partial solid phase fermentation of beet pulp with a probiotic association. The final product includes biomass of probiotic bacteria forming a biofilm on the surface of a phyto-carrier, products of their metabolism, phytosubstrate biotransformation products, prebiotics, pectins of beet, and phytocomponents. The bacterial composition of the preparations contains vegetative cells of three strains: *Bacillus subtillis, Bacillus licheniformis*, and a lactic acid bacteria complex. The ProStor preparation contains in the probiotic association a unique strain *Bacillus subtillis*—a producer of hydrolase class enzyme, which has anti-inflammatory and antiviral effects, stimulates the immune reactions of the body. A cellulolytic *Cellulomonas* microorganism is additionally introduced to the Ferm KM-1 probiotics composition and capable of both synthesizing enzymes that break down cellulose, and producing lysine, the essential amino acid. Depending on the type of fish and their food, the effect of biological action of bacteria varies. Therefore the preparation Ferm KM-1 increases the digestibility of all feed components, and, to the upmost degree, of fiber in case when the feed mix contains a lot of fiber, which is important, for example, for the carp. For the sturgeon on the protein diet, the preparation increases the digestibility and protein digestibility of feed.

Probiotic bacteria have an enhanced viability and resistance to adverse environmental conditions for they are in the form of a biofilm on a phytocarrier

(Figure 25.1). The preparations contain enzymes: cellulase, amylase, complex of proteases, lipase, as well as organic acids, biologically active substances, vitamins, amino acids, immunoactive peptides, and products of probiotics metabolism. The preparations comprise phytoparticles that are a cellulose microsorbent.

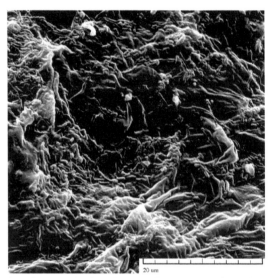

FIGURE 25.1   Microphotograph of the fermented sugar beet pulp with biofilm of probiotics

The preparations are featured with combining probiotics and prebiotics (mannans and glucans on cell walls of yeast *Saccharomyces cerevisiae*), and phytobiotics of the medicinal plants *Echinacea purpurea* and *holy thistle*. Echinacea has immunomodulatory properties. Echinacea preparations exhibit antibacterial, antiviral and antifungal properties. When intaking the Echinacea preparations at metabolic disorders, at the impact of different chemical compounds of toxic nature, contained in the feed (heavy metals, pesticides, insecticides, and fungicides), a stimulation of the immune system has been observed.

Holy thistle is used for prevention of various liver affections. Preparations of holy thistle increase protective properties of liver to infection and poisoning, stimulate the formation and excretion of bile. The positive effect of the plant also affects the liver, and the whole digestive tract.

The special feature of the ProStor and Ferm KM-1 products is the presence of yeast cell walls. They contain mannanoligosaccharides and beta-glucans, which effectively bond and absorb in the gastro-intestinal tract different pathogens. Beta-glucans have a stimulating effect and optimize the immune system.

The preparations increase the digestion and feed efficiency, growth rate, optimize the productive indices of fish, effective in the treatment and prevention of parasitic diseases.

The preparations that are hi-tech, bulk products of brown color, with slightly specific odor, which makes it easy to mix them with compound feed components. They tolerate forage production processes without loss of biological activity.

Warranty storage period of preparations is six months from the production date, subject to +30°C temperature and relative humidity up to 75 percent.

The ProStor and Ferm KM-1 preparations are used in the feed for the young and adult fish (the carp and the sturgeon). The preparation is administered in the feed in the feed mills or farms, by mixing. They are applied daily to feed on recommended zootechnical dosage rates (for the carp 1.0–1, 5 kg per ton of feed, for the sturgeon 1.5–2,0 kg per ton of feed). Side effects and complications at the use of preparations at the recommended doses have not been observed. There are no contraindications. Fish products after the use of preparations can be used without restrictions.

The efficacy of the ProStor preparation for fish is demonstrated in an experience with carp and sturgeon juveniles (Table 25.1). The preparation in an amount of 1.5 kg per ton of feed was introduced to the feed KM-2M for the carp, and in the amount 2.0 kg per ton of feed OT-7 for the sturgeon. The underyearlings were kept in aquaria in groups of 15 animals.

TABLE 25.1   The efficacy of the ProStor preparation for fish

Index	Carp		Bester (sterlet+beluga cross)	
	Experiment, 1.5 kg ProStor/t mixed feed KM-2M	Control, mixed feed KM-2M	Experiment, 2 kg ProStor/t mixed feed OT-7	Control, mixed feed OT-7
Absolute weight gain, g	8.9	6.25	14.2	6.6
% of control	142.4	100.0	215.1	100.0
Average 24-h weight grow rate, %	6.78	5.65	11.2	7.84
Food expenses, units	1.8	2.2	1.2	1.9
% of control	81.8	100.0	63.1	100.0
Survival rate, %	100	100	100	87

Fish breeding and biological indices of young carps and sturgeons as for absolute weight gain and average daily growth rate were higher than the ones of the control carp group, respectively, by 45 percent and 25 percent, and for control sturgeons—respectively, by 120 percent and 45 percent. The experimental sturgeon fry survival rate demonstrated was by 13 percent higher than the index of the control fish.

The cost of 1 kg of growth gain of the experiment carp was 23.4 rubles, which is 13 percent lower than in the controls (26.95 rubles). The cost of 1 kg of growth gain in the experiment *bester* equaled 21.2 rubles, which is 35 percent lower than in controls (33.0 rubles).

Feed cost indices (1.2 units) of pilot feed line with similar values of better feed foreign companies.

In experiments on the cultivation in a closed water supply for young sturgeons on the Ferm KM-1 diet at 0.1 percent in the production OT 7 feed of for young sturgeons, the *condition* factor, as well the absolute and average daily weight gain coefficient significantly increased (Table 25.2).

TABLE 25.2 Fish breeding and biological indices of growing for 2-year old sturgeon hybrids

Indices	Experiment Versions	
	Control	Experiment with *Cellulomonas*
Weight, g: initial	250.6±19.17	243.8±20.86
final	292.5±22.4	3042±31.2
Fullton's condition factor, %	0.35(100%)	0.39 (111%)
Absolute weight grow rate, g	41.9 (100%)	60.4(144%)
24 h grow rate, g	1.32 (100%)	1.95(148%)
24 average 24 h *grow rate*, %	0.50(100%)	0.72(144%)
Weight gain coefficient, unit	0.031(100%)	0.045(145%)
Food cofficient	1.2 (100%)	1.0(83%)
Survival rate, %	100	100

The results of checking the efficiency of the incorporation of the ProStor and the Ferm KM probiotic preparation to the mixed feed for the sturgeon demonstrate higher industrial productivity rates for *Russian-Siberian* sturgeon hybrid. The obtained data as fishery/biological indices allow to recommend the use of the ProStor and Ferm KM-1at the large-scale mixed fodder production for they provide higher productivity figures, lowering the costs for feed and stable health conditions for the fish cultivated.

KEYWORDS

- **Biofilm**
- **Feed**
- **Fish farming**
- **Phytobiotics**
- **Probiotics**

REFERENCES

1. Bychkova, L. I.; Yukhiмenко, L. N.; Khodaκ et al.; A. G.: Fish farming, **2008,** 2, 48 (in Russian).
2. Pokhilenκo, V. D.; and Perelygin, V. V.; News of medicine and pharmacy. **2008** *8*(259), 56, (in Russian).
3. *Harbarth, S.; and Samore,* M. H.; *Emerg. Infect. Dis.* **2005**, *11*, 794.
4. Pickering, A. D.; Stress and Fish. A.D. Pickering (ed.). London-N.Y.: Acad. Press, **1993**; 1.
5. Matsuzaki, T.; *Immunol. Cell. Biol.* **2000**, *78*(1), 67.
6. GROZESKU, YU. N.; BAKHAREVA, A. A.; SHULGA, E. A.; News Bulletin of Samara Scientific Center, RAS, 11, 1(2). **2009**, 42, (in Russian).
7. Sariev, B. T.; Tumenov, A. N.; Bakaneva, YU. M.; and Bolonina, N. V. *ASTU News Bulletin. Ser. Fish Farming.* **2011**, *2*, 118, (in Russian).
8. Panasenкo, V. V.; *Fish Farming.* **2008**, *1*, 74, (in Russian).
9. Ponomaryov, S. V.; Grozesku, YU. N.; and Bakhareva, A. A.; Industrial fish farming. Moscow: Kolos, 2006. 320 p. (in Russian).

CHAPTER 26

A CASE STUDY ON DEVELOPMENT OF A NEW AEROBIC-ANAEROBIC BIOREMEDIATION TECHNOLOGY

SERGEY GAYDAMAKA[*] and
VALENTINA P. MURYGINA LOMONOSOV[**]

Department Chemical Enzymology, Chemistry Faculty, Moscow State University, Leninskye Gory, 1, build.11, Moscow,119992, Russia; [*]E-mail: s.gaidamaka@gmail.com; [**]E-mail: vp_murygina@mail.ru, vpm@enzyme.chem.msu.ru

CONTENTS

26.1   INTRODUCTION

The main oil production areas in Russia are situated in the Northern Siberia, and in the same places there are situated most extensive bogs polluted with oil. Application of remediation technologies, developed in Russia, on impassable bogs polluted with oil is almost impossible technically and economically unfavorable. Besides a severe climate with cold and long winters and short cool summers it is caused also by absence of any roads in tundra and forest-tundra and emergency oil spills on fenny bogs impassable for special machinery devices.

Therefore an elimination of such spills and their consequences on the bogs is a very actual and difficult problem there. Depth of oil penetration on bogs doesn't exceed of 0.6–1.0 m and often is propped up with water or permafrost. Processes of self-restoration such bogs can prolong for several hundred years. The pollution can extend on width there and the irreparable damage will be caused to the Nature of the Polar Region.

In 2011 there was an attempt to clean from oil a strongly polluted bog with using (augmentation) of a bacterial oil-oxidizing preparation Rhoder. Oil spill was spring, and oil was partially collected with a pump for sludge. Three times the bog was watered with the Rhoder and one time with a fertilizer and lime. As a result level of oil pollution in the peat was decreased by 32 percent to 98 percent depending on initial concentration of oil which varied from 21 to 29 kg of crude oil on 1 kg of absolutely dry matter (DM) to 450–850 g/kg DM, and a depth of penetration of oil into the moss. The received results have induced a development of a new remediation technology in laboratory conditions with using of electron acceptors and the Rhoder to enhance of oil oxidation on the surface and in the depth of the peat. In this paper there is presented an attempt to develop a new aerobic-anaerobic bioremediation technology for fenny bogs polluted with oil for using it in the northern part of Russia.

26.2   MATERIALS AND METHODS

The microbial oil-oxidizing preparation Rhoder was used in laboratory experiment. The Rhoder consists of two bacterial strains Rhodococcus (R. ruber Ac-1513 D and R. erythropolis Ac-1514 D) picked out from soils, polluted with oil. Strains were not pathogenic for people, animals and plants and also don't cause mutations in bacteria. The Rhoder is allowed for broad use in the nature. It was successfully applied to bioremediation of oil sludge, soils, bogs and surfaces of water from oil pollution [1–8].

Installation, which was made from vertical plastic pipes (5 models) with a length of 100 cm and diameter of 10 cm, was attached to a board (Figure 26.1). In each model two openings with a diameter of 2 cm at distance of 40cm and of 90cm from the upper edge were made for sampling. Each model was filled with the natural peat polluted with oil with a high concentration of hydrocarbons (HC) from 370 g/kg to 550 g/kg of DM.

26.3   SCHEME OF THE EXPERIMENT

- Model No. 1—negative control in which was added water for maintenance of high humidity of the peat, which was typical for bogs.
- Model No. 2—activation of indigenous microorganisms with mineral fertilizers that were added into the top layer of the peat into the depth of 10 cm, and introduction of a gaseous electron acceptor into the depth of 40 cm from the top layer of the peat in the model.
- Model No. 3—processing of the top layer of the peat into the depth of 10 cm with the water solution of the Rhoder and fertilizers and injection of the gaseous electron acceptor into the depth of 40 cm from the top layer of the peat in the model.
- Model No. 4—processing of the top layer of peat into the depth of 10 cm with the water solution of the Rhoder and fertilizers and liquid electron acceptor into the top layer of the peat in the model.
- Model No. 5—processing of the top layer of the peat into the depth of 10 cm with the water solution of the Rhoder and fertilizers and injection of the liquid electron acceptor into the depth of 40 cm from the top layer of the peat in the model.

26.3.1   CARRYING OUT BIOREMEDIATION

Soils in the models No 3–5 were processed three times with working solution of the Rhoder with a number of hydrocarbon oxidizing cells (HCO) of $1.0*10^8$ cells/ml by watering with an interval in 3 weeks. The fertilizer («Azofoska» C:N:P 16:16:16) was used, and 40 ml of the solution was added to the models three times. The gaseous and liquid electron acceptors were used and injected into models, according to the scheme of the experiment. The top layers of the peat in all models were maintained humidity not less than 60%, and the top layers of the peat were mixed two times a week and before each introduction of fertilizer and the Rhoder.

26.3.2   SAMPLING

Soils sampling from models were made before the experiment beginning and before every application of the fertilizer and the Rhoder and each injection of electron acceptors which were entered into models. Samples from models were selected from the depth of 0-10 cm, 40 cm and 90 cm from the upper edge of each model for conducting of chemical, agrochemical and microbiological analyses.

26.3.3   CHEMICAL AND AGROCHEMICAL ANALYSES

Oil in each sample of the dry peat was extracted on a Sockslet device with boiling $CHCl_3$, and gravimetrically determined. Then each dry material extracted by chloroform was fractioned on a minicolumn with silica gel (Diapak-C). Oil products were analyzed by the gas chromatograph (GC). GC model is the KristalLuks 4000м (by company Metakhrom) with the NetChrom V2.1 program, the column OV-101 length of 50 m, internal diameter of 0,22 mm, thickness of the phase of 0,50 microns, the FID detector, the temperature of the detector 300°C, the evaporator temperature of 280°C, the gradient from 80°C to 270°C, the velocity of raising temperature was 12°C per minute. [9].

pH of each sample, humidity, and the general maintenance of the available nitrogen and phosphorus were determined with colorimetric methods [10].

26.3.4   MICROBIOLOGICAL ANALYSES

MPN of microorganisms was determined by using tenfold dilutions and cultivations on meat-peptone agar in Petri dishes and using of selective agar nutrients for identification of ammonifying microorganisms, actinomicetes, pseudomonas, oligotrophic bacteria, and micromycetes. MPN of anaerobic microorganisms (first of all SRB) in samples of the peat which have been selected from the depth of 40см and 90 cm from models were determined on the liquid Postgate's medium [11].

Determination of MPN of oil-oxidizing microorganisms in samples of peat were used the modified liquid Raymond's media with oil as a sole carbon source (g/l): Na_2CO_3 - 0.1; $CaCl_2*6 H_2O$—0.01; $MnSO_4*7 H_2O$—0.02; $FeSO_4$—0.01; $Na_2HPO_4*12H_2O$ -1.0; KH_2PO_4—1.0; $MgSO_4*7 H_2O$—0.2; NH_4Cl—2.0; $NaCl$—5.0; pH = 7.0 [12].

26.3.5   RESULTS AND DISCUSSION

Laboratory experiment was performed on the models which imitated of an over wetted bog polluted with oil. Preliminary microbiological analyses of samples taken from the top, middle and bottom layers of the peat on the length of models showed that the peat in all models had different species of microorganisms: Bacillus, Pseudomonas, Rhodococcus, SRB, and Penicillium. In the top layers of the peat (0–10 cm) in each model there were discovered the MPN of hetero-trophic bacteria (HT) from $6.0*10^7$ to $1.1*10^8$ CFU/g of the peat, HCO bacteria from $9.1*10^5$ to $9.4*10^6$ cells/g of the peat. In the top layers of the peat in mod-els anaerobic microorganisms didn't determine. In samples of the peat, which have been selected from the middle parts of models, the MPN of HT varied from $8.1*10^5$ to $3.7*10^7$ CFU /g of the peat, MPN of HCO bacteria varied from $8.2*10^2$ to $9.8*10^4$ cells/g of the peat. MPN of anaerobic microorganisms (SRB) $1.0*10^2$ cells/g of peat were found in the samples from the middle of models No 3 and 4. In the bottom samples of the peat in these models there were found an-aerobic and microaerophilic bacteria with MPN from $2.1*10^6$ to $4.9*10^7$ CFU/g of the peat and HCO bacteria from $7.1*10^3$ to $1.0*10^6$ cells/g of the peat. SRB were found in the bottom samples from the models No 2, 3 and 5 with the MPN from $1.0*10^2$ to $1.0*10^5$ cells/g of the peat (Table 27.1).

Agrochemical analyses show that the content of nitrogen and phosphorus in the peat in models was rather high and the ratio of C:N:P was in average 100:0.1:0.01 (Table 26.1).

TABLE 26.1 Microbiological and agrochemical characteristics of peat samples from the different length of the models before bioremediation

№ model	Point of sampling	pH	HT CFU/g of peat	HCO, cells/g of peat	SRB, cells/ml	$N-NH_4^+$ mg/kg of peat	PO_4^{3-} mg/kg of peat
1	Top	5.9	$8.5*10^7$	$9.4*10^6$	–	507.13	459.7
	Middle	6.2	$8.1*10^5$	$8.1*10^2$	0	516.2	393.6
	Bottom	6.7	$2.9*10^6$	$9.7*10^4$	0	544.1	418.3
2	Top	5.8	$8.3*10^7$	$9.4*10^6$	–	450.8	393.6
	Middle	6.4	$3.6*10^7$	$9.8*10^4$	0	453.7	471.1
	Bottom	6.1	$1.4*10^7$	$7.1*10^3$	$1,0*10^5$	495.2	318.6
3	Top	6.0	$1.1*10^8$	$1.1*10^5$	-	427.3	402.6
	Middle	6.2	$3.7*10^7$	$8.7*10^4$	$1,0*10^2$	440.7	401.1
	Bottom	6.3	$4.9*10^7$	$1.1*10^4$	$1,0*10^2$	428.8	370.2
4	Top	6.3	$1.1*10^8$	$9.1*10^4$	-	402.5	391.2
	Middle	6.3	$6.9*10^6$	$1.1*10^4$	$1,0*10^2$	448.6	411.9
	Bottom	6.1	$2.8*10^6$	$9.4*10^4$	0	463.9	506.7
5	Top	6.0	$6.0*10^7$	$9.5*10^5$	-	494.7	329.7
	Middle	6.2	$3.4*10^7$	$1.1*10^4$	0	454.1	278.1
	Bottom	5.9	$2.1*10^6$	$1.0*10^6$	$1,0*10^5$	454.7	364.2

Note: –not detection

After completion of the experiment the MPN of microorganisms (HT and HCO bacteria) grew on 1 or 2–3 orders practically in all models in the top layers of the peat and decreased on 1-2 orders in the middle and bottom layers of the peat. At the same time the number of anaerobic microorganisms including SRB in all models significantly grew in the middle and the bottom layers of the peat that can be connected with formation of own biocenosis in each model (Table 26.2).

TABLE 26.2 Microbiological and agrochemical characteristics of peat samples from the different length of the models after the end of the bioremediation

№ model	Point of sampling	pH	HT CFU/g of peat	HCO, cells/g of peat	Other anaerobic bacteria/ SRB, cells/ ml	$N-NH_4^+$ mg/kg of peat	PO_4^{3-} mg/kg of peat
1	Top	5.8	$5.7*10^7$	$1.1*10^7$	-	316.9	168.9
	Middle	5.7	$2.6*10^8$	$9.9*10^1$	$0/5*10^3$	425.8	276.7
	Bottom	5.2	$3.4*10^7$	$8.2*10^3$	$3.1*10^8/$ $5*10^3$	417.4	226.6
2	Top	6.3	$2.1*10^9$	$8.6*10^7$	-	493.7	429.5
	Middle	5.9	$5.0*10^7$	$1.0*10^3$	$2.0*10^8/$ $5*10^3$	297.9	273.5
	Bottom	6.5	$8.0*10^7$	$6.1*10^6$	$0/1.8*10^4$	215.3	204.6
3	Top	6.4	$1.0*10^9$	$8.8*10^7$	-	589.9	173.5
	Middle	6.6	$4.9*10^6$	$1.3*10^4$	$1.2*10^7/$ $5*10^3$	589.4	369.9
	Bottom	6.3	$4.7*10^6$	$1.1*10^6$	$7.4*10^7/$ $5*10^3$	258.0	195.3
4	Top	6.8	$2.2*10^8$	$1.0*10^8$	-	229.1	251.3
	Middle	6.4	$8.9*10^5$	$8.1*10^3$	$6.6*10^7/$ $5*10^3$	214.0	203.2
	Bottom	6.5	$1.2*10^6$	$8.5*10^3$	$5.0*10^7/$ $1.8*10^4$	290.5	352.2
5	Top	6.5	$4.3*10^8$	$1.0*10^8$	-	122.5	349.8
	Middle	5.6	$1.7*10^6$	$7.6*10^4$	$1.5*10^7/$ $1.8*10^4$	313.3	142.9
	Bottom	6.2	$1.2*10^6$	$1.2*10^3$	$6.6*10^7/$ $5*10^3$	294.7	199.2

Note: –not detection

Concentration of biogenic elements in all layers of models changed and even decreased that can be connected with activation of microorganisms in models. In the bottom and in the middle parts of the models concentration of biogenic elements decreased that can be connected with activation of anaerobic microorganisms in these models (Table 26.2).

Chemical analyses have shown that the initial concentration of oil in the models have varied on height of the models from 370 g/kg DM to 550 g/kg DM. The concentration of oil in the models has decreased according the gravimetric analyses on the average by 24–34 percent, including by 19 percent in the control model after finishing this experiment.

GC analyses of oil products from models on their length are presented in Figures 26.1 to 26.5.

Results of GC of the analyses show that in control model there are processes degradation of oil in the top part. The quantity of peaks practically doesn't change, but their height and areas decreases (Figure 26.2). The amount of oil products in the control model decreased in the top layer by 54 percent. In the middle of the model the number of peaks decreased by 1 peak, but their area increased. In the bottom of the model the quantity of peaks increased by 1 peak, but the area of peaks and their height (Figure 26.2) significantly increased probably at the expense of an oil filtration down. GC analyses of oil products on length of the model No. 2 are presented in Figure 26.3.

FIGURE 26.1 The laboratory installation, modeling a fenny bog, for development of a new technology of bioremediation of peat, polluted with oil

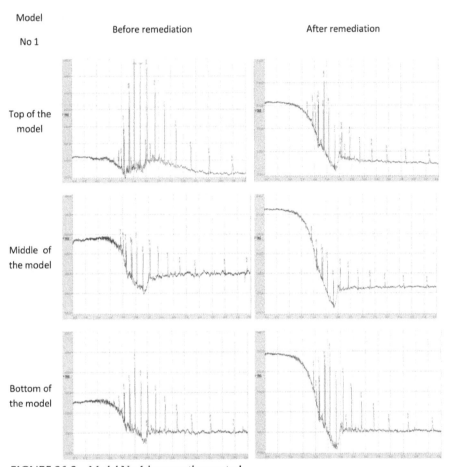

FIGURE 26.2   Model No 1 is a negative control.

Results of GC analyses (Figure 26.3) have shown that processes of oil degradation in the top, middle and bottom parts of this model took place. In the top and the middle parts of this model the quantity of peaks and their areas have decreased. In the bottom part of the model the quantity of peaks doesn't change, but their areas slightly have decreased. The concentration of oil products in the model decreased in the top part by 74 percent, in the middle part by 24 percent and in bottom by 5 percent. In this model the gaseous electron acceptor promotes degradation of oil products by anaerobic microorganisms in the middle and bottom parts. But in the top of the model aerobic indigenous microorganisms have worked. GC analysis of oil products in the model No. 3 is presented in Figure 26.4.

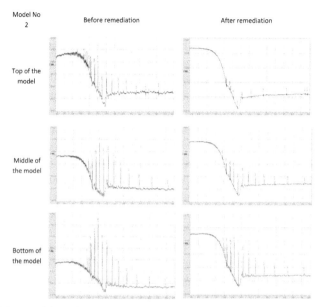

FIGURE 26.3   Model No 2. Activation of indigenous microorganisms with fertilizers and injections of the gaseous electron acceptor into the middle part of the model.

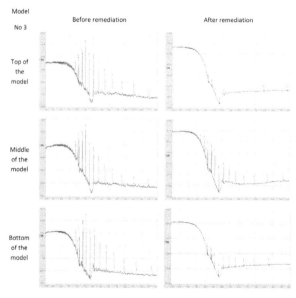

FIGURE 26.4   Model No. 3. Augmentation with the Rhoder and fertilizers and injections of the gaseous acceptor of electrons into the middle part of the model.

Results of the GC analyses (Figure 26.4) have shown that in the model No 3, in which the Rhoder with the MPN of HCO bacteria $1.0*10^8$cells/ml and fertilizers were added three times and also the gaseous acceptor of electrons also were injected three times, have activated processes of oil degradation in the top and bottom parts of the model. In the middle part of the model, the quantities of peaks and areas have increased. In the top part of the model the concentration of oil products have decreased by 88 percent, in bottom part by 68 percent. In the middle part of the model GC analyses have not shown decrease in oil products. The acceptor of electrons probably has promoted degradation of oil products by anaerobic microorganisms in the bottom part of the model.

GC the analyses of oil products in the model No. 4 are presented in Figure 26.5.

Results of GC analyses have shown (Figure 26.5) that addition of the Rhoder, fertilizers, and the liquid acceptor of electrons into the top layer of the model in three times has decreased the quantity of peaks from 15 to 6 and significantly has decreased its areas there. 1). In the middle and the bottom parts of the model the quantity of peaks has not changed, but the areas of peaks in the bottom part of the model have decreased (Figure 26.5). In this model the concentration of oil products have decreased in the top part almost by 96 percent, in the bottom part by 27 percent. In the middle part of the model the areas of peaks have increased by 24 percent by the end of the experiment. Probably the acceptor of electrons added into the top layer of the peat is not so good promoter for degradation of oil products by anaerobic microorganisms.

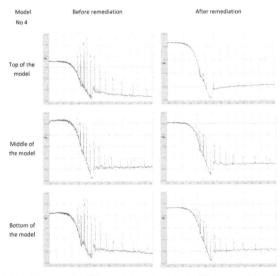

FIGURE 26.5   Model No. 4. Addition of the Rhoder, fertilizers and the liquid acceptor of electrons into the top layer of the model three times.

The GC analyses of oil products in the model No. 5 are presented in Figure 27.6. Results of GC analysis have shown that in the top part of the model (Figure 26.6) the quantity of peaks has decreased from 18 to 10 and very significantly its total area has decreased. The concentration of oil products has decreased in the top layer of the peat by 99 percent. In the middle part of the model the quantity of peaks doesn't change, but the total area of peaks has decreased. And concentration of oil products has decreased on the average by 64 percent. In the bottom part of the model the quantity of peaks has decreased from 19 to 15. The liquid acceptor of electrons, injected into the middle part of the model, significantly promotes degradation of oil products by anaerobic microorganisms in the middle and bottom parts in this model. In the top part of the model the process of oil degradation has provided with the Rhoder.

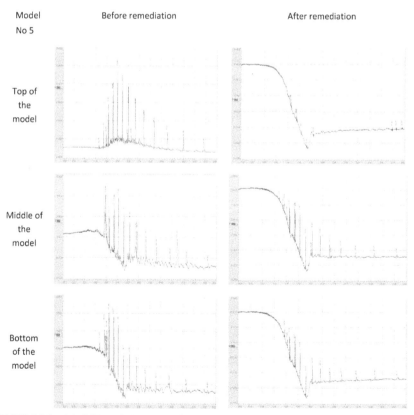

FIGURE 26.6   Model No. 5. Augmentation with the Rhoder and fertilizers and injection of the liquid acceptor of electrons into the middle part of the model three times.

Thus, the received results show that the liquid electron acceptor in comparison with the gaseous electron acceptor has more good effect on the degradation of oil. And it is more expedient to inject the liquid electron acceptor into the middle part of the model (into the depth of 40 cm). These results are the first step to develop of a new aerobic and anaerobic bioremediation technology for strongly polluted fenny and almost impassable bogs in the North of Russia. Because there is impossible to collect completely spills of oil and perform a classic remediation technology on polluted bogs especially with using of specialized equipment and devices. It should be noted that the liquid acceptor of electrons has no relation to iron salts; this acceptor is eco-friendly and well makes activation of anaerobic indigenous microorganisms.

CONCLUSIONS

The obtained results showed that both studied acceptors of electrons well work into an anaerobic zone. Into the aerobic zone the Rhoder works more effectively in comparison with indigenous microorganisms. The oil-oxidizing effect of the Rhoder in combination with the gaseous or the liquid acceptor of electrons showed good results. However augmentation with the Rhoder and fertilizer and the liquid electron acceptor, injected into the middle part of the model, had the best effect on oxidizing of oil there.

KEYWORDS

- **Acceptor of electrons**
- **Augmentation**
- **Microorganisms**
- **Model**
- **Oil**
- **Peat**

REFERENCE

1. Murygina, V.P.; Arinbasarov, M. U.; and Kalyuzhnyi, S.V.; Ecology and Industry of Russia. **1999** (*8*), 16. (in Russian)
2. Murygina V., Arinbasarov M., and Kalyuzhnyi S.: *Biodegradation.* **2000** (*11*), *6*, 385. (in Russian).
3. Murygina, Valentin, A. P.; Markarova, Maria Y.; Kalyuzhnyi, Sergey V; *Environ. Inter,* **2005,** *31* (2), 163.

4. Ouyang, W.; H.Liu, Y.yu; Murygina, V.;.Kalyuzhnyi, S.; and Xiu, Z;. *Process Biochem.* **2005**, *40*(12), 3763.

5. Wei Ouyang; Hong Liu; Yong-Yong Yu; Murygina, V.; Kalyuzhnyi, S.; and Zeng-De Xiu; Huanjing Kexue/*Environ. Sci.* **2006**, *27*(1), 160.

6. De-Qing, S.; Jian, Z.; Zhao-Long, G.; Jian, D.; Tian-Li, W.; Murygina, V.; and Kalyuzhnyi, S.; *Water. Air. Soil Polluti.* **2007**, *185* (1–4), 177.

7. Murygina, Valentina, Markarova, Sergey, Maria, Kalyuzhnyi: In Proc. of IPY-OSC Symp., Norway, Oslo. **2010**. http://www.ipy-osc.no/

8. Murygina, V.; Gaidamaka, S.; Iankevich, M.; and Tumasyanz, A.; *Progress. Environ. Sci. Technol.* **2011**, III, *791.*

9. Drugov, Yu.S.; Zenkevich, I. G.; and Rodin, A. A.; Gas chromatography Identification of Air, Water and Soil and Bio-nutrients Pollutants. Binom, Moscow **2005**, 752 p (in Russian).

10. Mineev, V. G. (Ed.) Practical handbook on Agro chemistry. Moscow State University, Moscow, Russia, **2001**. 688 p (in Russian).

11. NETRUSOV, A. I.; (Ed.) Practical Handbook on Microbiology. Academia, Moscow, Russia. , **2005**, 608 p (in Russian).

12. Nazina, T.; Rozanova, YE.; Belyayev, S.; and Ivanov, M.; Chemical and microbiological research methods for reservoir liquids and cores of oil fields. Preprint Biological Centre Press, Pushchino, **1988**. 35 p (in Russian).

CHAPTER 27

DEVELOPMENT OF NONTOXIC METHODS OF RODENT POPULATION CONTROL AS AN ALTERNATIVE APPROACH FOR BIG CITIES

V. V. VOZNESSENSKAYA and T. V. MALANINA

A. N. Severtzov Institute of Ecology & Evolution, 33 Leninski prospect, Moscow, 119071, Russia, email: veravoznessenskaya@gmail.com

CONTENTS

27.1 INTRODUCTION

Rodents cause considerable economic damage to field and fruit crops on annual basis.

In the urban area in addition to economic losses, human health, and safety from rodent transmissible zoonoses are of concern. Highly toxic methods are applied currently in Russia to manage rodent populations, which are not safe for humans and other mammalian species [22]. Major pitfalls of current approaches: high toxicity to humans and other nontarget species; environmental pollution; and development of avoidance behavior and rodenticide resistance in rodents. Moreover, individuals, survived after rodenticide treatment exhibit reproductive outbreaks [13]. Zoonoses attributable to rodents are exacerbated during periods where their population erupts. Methods that can dampen these irruptive population cycles would prove highly desirable. It is our goal to develop a product that will dampen the amplitude of these rodent population cycles. Rodent population size is regulated by external as well as internal (zoosocial) factors. Predator population density is the most influential external factor [6]. Major internal regulating factor is the rodent population density itself. Our investigation is aimed to develop a product based on a number of substances involved in the regulation of rodent population density under natural conditions. Reproductive control in wildlife species who comes into conflict with humans has received increasing attention as a humane method for managing wildlife populations. Moreover, modeling studies show that contraception as a tool to management populations is best suited for species with high population turnover (i.e., short generation time and high reproductive output). Thus, rodents are ideal targets for this management tactic. Predator urine is used as a wildlife management tool to repel herbivorous animals from areas. Fundamentally, avoidance of predator urine by potential prey, and by implication the areas where predators frequent, is presumably evolutionary advantageous because it lowers the risk of predation. Potential prey can discriminate predator urines as opposed to that of other herbivores on the basis of the urine's odor. One consequence of a high meat diet is the presence of sulfurous compounds in the urine. These compounds result from protein digestion and metabolism. When sulfurous compounds are removed from predator urines by mercury treatment herbivorous animals are no longer repelled by the urine's odor [12]. Our previous research showed the effects of predator odors on various aspects of rodent reproductive behavior and reproductive output [16, 16, 8]. Felinine is a unique sulfur-containing amino acid found in the urine of domestic cats [14]. Felinine is unstable in water solution and exists in the form of mixture of amino acid and sulfur-containing volatile compounds. One of four of them: 3-mercapto-3-methyl-1-butanol has a characteristic cat odor and believed to be a pheromone. Miyazaki et al., 2006). We now present evidence to support bioactivity of L-felinine with rodents.

27.2 MATERIALS AND METHODS

27.2.1 TEST SUBJECTS

Test subjects were 3–4 month old Norway rats (Rattus norvegicus) and 4–6 month old mice (Mus musculus); both from an outbred laboratory population. Before the start of the experiments, females were housed in groups of 3–4, and males were housed singly. Experimental rooms were illuminated on 14:10 hrs light:dark schedule, and maintained at 20–22°C. Food and tap water were provided *at libitum*. Virgin females in proestrus/estrus, as determined by vaginal cytology, were chosen for the mating experiments. Sexually experienced males that were not mated in the 14 days before the test were used as sires. The morning after pairing, the females were checked for successful mating, as indicated by the presence of a vaginal plug. Successfully mated females were then housed singly.

27.2.2 REPRODUCTIVE OUTPUT

For each experimental group, the total number of offspring was counted as well as number of pups per female; sex ratio was determined. In addition, we also weighed pups at the time of weaning (21-st day after birth).

27.2.3 COLLECTION OF URINE.

Urine from domestic cats (*Felis catus*) was used as a source of predator chemical cues. These cats normally hunt for mice and have mice as part of their diet. If needed, additional meat was added to their diet. Freshly voided urine was frozen (–22°C). Once defrosted, urine was used only once. Nonpredator urine was obtained from guinea pigs. Individuals of these species were placed into metabolic stainless-steel cages overnight, and urine was collected and stored using the method described above. Urine was collected and stored at –22°C.

27.2.4 "OPEN FIELD" TEST WITH ADDED STRESS (ROUGH HANDLING CONDITIONS)

An "open arena" (D = 0.7 m) with bright lights was used. Pregnant females were placed for 15 min in the center of the arena on 1-st, 3-d, 5-th and 7-th day of gestation. During the test, we also used a buzzer, which made a loud noise, every 5 minutes. In addition, mice were handled roughly to physically induce stress. Blood samples from sublingual vein were drawn after each test for progesterone and corticosterone assay.

27.2.5 ASSAY FOR PROGESTERONE AND CORTICOSTERONE

Animals within each treatment were randomly assigned to one of four cohorts. Blood samples (100 µl) were obtained from sublingual vein every second day for each cohort for each of the treatment for the first 7 days of gestation. This minimizes the handling and sampling of individual mice, while allowing a detailed study of changes in hormonal pattern as a function of time and treatment. Our experience shows that this method of repeated blood sampling has no long-term effect on visible scarring associated with traditional tail sampling technologies. Samples were centrifuged and the plasma frozen at $-20°C$ until subsequent analysis. Plasma progesterone and corticosterone were assayed (in duplicate) by enzyme immunoassay (EIA) method (DRG, USA).

27.2.6 ASSAY FOR FECAL CORTICOSTERONE METABOLITES

In small animals like mice, the monitoring of endocrine functions over time is constrained seriously by the adverse effects of blood sampling. Therefore, we used noninvasive technique to monitor glucocorticoids with recently established 5a-pregnane-3ß,11-ethol,21-triol-20-one EIA [17] to assess adrenal activity in mice under conditions of long lasting exposures to predator odors. Mice were exposed to L-Feinine (0.05%) on everyday basis for period of two weeks. On completion of exposures fecal material was collected from each animal over 24 hrs. Extraction procedure was performed with 80 percent methanol. Concentration of corticosterone metabolites was measured with spectrophotometer (Spectramax340, Molecular Devices, USA) at 450 and 670 nm. Specific antibodies were received from prof. E.Möstl laboratory (University of Veterinary Medicine, Vienna).

27.2.7 IMMUNOHISTOCHEMISTRY ASSAY

To visualize activated neurons on olfactory bulbs sections in response to stimulation, Fos protein immunohistochemistry was used [4]. Fos protein is a product of c-fos known as immediate early gene which is induced quickly by different stimuli including cell depolarization [15]. Labeling Fos provides a physiological marker of neurons activated in response to specific stimuli. Half life span of protein Fos is 2 hrs: depending on specific characteristics and neural cell localization optimal exposure time for maximal Fos detection may range from 45 to 90 min [21]. To stimulate main and accessory olfactory system mice were exposed L-Felinine (0.05% in water) for 40 min using half duty cycle (1 min—specific odor, one minute—clean air). Immediately after exposure mice were perfused with 3 percent paraformaldehyde in phosphate buffer. Olfactory bulbs were removed and postfixed in paraformaldehyde for 16hours. We used standard procedure for fixation of olfactory bulbs, cryoprotection and immuno-histochemical staining of olfactory bulbs sections (DellaCorte, 1995). We used indirect avidin/biotin method; horseradish peroxidase was used as enzymatic label, diaminobenzidin (DAB) was used as chromogen. Sections were made at 20 μm using cryostat Triangle Biomedical. Immunostaining was made according to standard three day protocol using primary antibodies Santa Cruz Biotechnology (USA):c-fos (4) sc-52, dilution 1: 500. For visualization and counting of Fos positive cells we used Nikon©Eclipse E400 microscope with camera Nikon©Coolpix 990. For picture analyses we used ImageJ (NIH).

27.2.7 EXPERIMENTAL DESIGN

The experimental method consisted of applying 0.2 ml of a test solution (urine of 0.05% L-felinine) to the bedding of pregnant rats or mice every other day for different time durations. This application maximized the likelihood of physical and odor exposure of the test stimulus to the female. In experiments, three treatment levels were used:

(1) tap water (WAT), as a negative control;

(2) urine from guinea pigs maintained on a vegetarian diet (vegetables, grains and water *ad libitum*), as a urine control (GPU);

(3) urine from domestic cats maintained on a feral mouse diet (CU), as a model stimulus representing unadultered predator urine. Cats were maintained on the feral mouse diet for 14 days before urine collection;

After mating, females were randomly assigned to treatment groups: WAT, GPU, CU. Mean differences among treatment groups were determined in separate analyses for the number of pups and sex ratios using software STATISTICA8..

27.3 RESULTS

Exposure of mated female mice to intact cat urine provoked block of pregnancy in 30-70 percent of cases depending on the season. At the same time average percent of pregnancy block in control animals did not exceed 15 percent ($n = 16$, p<0.001, Fisher test). In autumn-winter season exposure to L-felinine (0.05%) provoked pregnancy block in 70 percent of mated female mice while in control group we observed only 20 percent of females with block of pregnancy (n = 20, p = 0.043, Fisher test). In spring-summer analogous exposures to L-felinine provoked pregnancy block in 62.5 percent of mated females compared with 12.5 percent in control (n = 9, p = 0.046, Fisher test). In felinine treatment group (Figure 27.1) number of pups per fertile female was 2.5 ± 1 while in con-trol–5.70 ± 1.00 (n = 28, p = 0.046 Mann-Whitney U test). Sex ratios in mice also were affected in favor of males by both treatments: cat urine (p<0.001) and L-felinine (p<0.01). Data presented in Figures 27.2 and 27.3. Exposure of preg-nant rats to L-felinine did not affect significanty litter size though we observed significant reduction in cat urine treatment group. On the contrary sex ratio in rats was affected in both treatment groups in favor of males: urine (p = 0.0007), L-felinine (p = 0.0007).

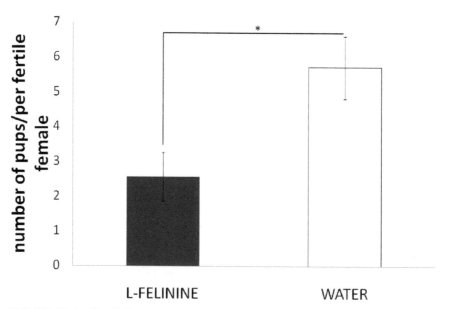

FIGURE 27.1   The influence of exposures to L-felinine (0.05%) during gestation on reproductive output in house mouse *Mus musculus* (Mann-Whitney U Test *p≤0, 05, n=28, ⊤ - SEM)

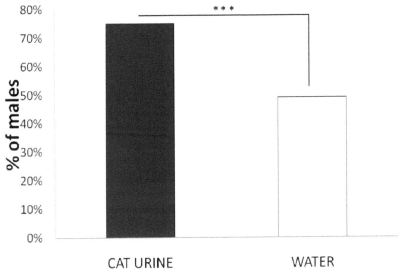

FIGURE 27.2   The influence of the cat urine *Felis catus* exposures during gestation on sex ratio in house mouse *Mus musculus* (***p≤0,001, n (cat urine) = 52, n(water) = 118, Fisher test).

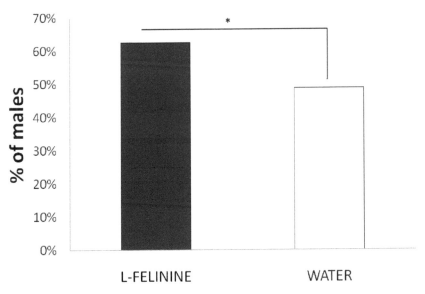

FIGURE 27.3   The influence of the L-felinine (0.05%) exposures during gestation on sex ratio in house mouse *Mus musculus* (*p≤0,01, n(L-felinine)=72; n(water) = 160, Fisher test).

We observed clear elevation of plasma corticosterone ($p<0.001$, n = 8, Tukey test) in response to felinine in mice (Table 27.1). As positive control we used "open arena" test with "added stress". Mice responded to this kind of treatment with elevated corticosterone but we observed habituation during the course of consecutive placements (days 1–5). At the same time mice did not habituate to consecutive exposures to felinine. We also observed such a habituation in mice introduced to other novel stimulus–guinea pig urine. To explore for how long predator chemical cues may provoke elevated corticosterone we exposed mice to L-felinine for 2 weeks. On completion of exposures fecal glucocorticoid metabolites were measured for each animal. In control group concentration of corticosterone metabolites was 203, 85 ± 47, 74 ng/ml, in felinine treatment group–702, 15 ± 122, 24 ng/ml (n = 13, $p<0.001$, t-test). The response of laboratory naive animals to predator scents and failure to habituate to the stimulus indicate the innate nature of the response. Chronically elevated cortocosterone may be responsible for the induction of pregnancy block.

TABLE 27.1   The influence of exposures to cat urine *Felis catus* on plasma corticosterone in house mouse *Mus musculus*. (M±SD; n = 8, each group)

	Plasma corticosterone (ng/ml)		
	1 -st day	3-d day	5-th day
Cat urine	681,25±135,16	706,25±123,63	716,25±105,55
Open field with "added stress"	371,25±175,05	183,75±86,34	96,25±34,61
Guinea pig urine	278,87±96,91	204±26,98	168,75±25,87
Water	92,75±43,51	77,88±22,8	84,25±17,7

We did not observe any differences in plasma progesterone for cat urine/felinine treatment groups and control animals. Immunohistochemical studies revealed neural activation in response to stimulation with L-felinine at the level of main olfactory bulb as well as at the level of accessory olfactory bulb indicating the involvement of both systems (main olfactory and vomeronasal) in detection of L-felinine which is important if practical applications are considered. In solution L-felinine is unstable; exists in form of mixture of amino acid and sulfur-containing volatile compounds. Most likely that 3-mercapto-3-methyl-1-butanol (felinine derivate) binds to receptors in main olfactory epithelium.

27.4 DISCUSSION AND CONCLUSIONS

Reproductive traits in rodents are affected by a number of environmental, social and chemosensory factors, e.g. the nutritional status of females will influence ovulation rate and litter size [5], as will exposure of females to other rodents of various social status [7]. Other well-described influences include synchronization of ovulation amongst female cohorts [24], acceleration or delay of puberty [18], [10], pregnancy block owing to stress, and failure to implant blastocysts when female rodents are exposed to the odor or urine of strange males [2].

The majority of these studies on reproductive inhibition have focused on intraspecific influences of semiochemicals and how they influence reproductive output and behavior in females. A few studies have focused on between-strain influences or interspecific influence, although the source odor generally is still confined to rodents.

During our investigation on the effects of predator odor on rodent reproduction and repellency, we found that female rats exposed to cat urine during pregnancy had reduced litter sizes at parturition [22]. Exposure to predator odor also caused disruptions of the oestrous cycle [16]. These effects bear striking similarities to the studies of the effects of rodent urine odor on intraspecific rodent reproduction. If such similarities are broadly based, then similarities in mechanisms of perception, reproductive physiology, and chemical nature of stimulus might be anticipated.

We do not believe that reduction in litter size is attributable to an adaptive response by rodents to predator odors. Rather, we propose the following interpretation. Urine contains information about the identity of individuals, reproductive status, and dominance status. We postulate that urine also contains information about environmental quality as reflected by nutritional status. Investigation of urine from a variety of sources would serve as an efficient way to integrate environmental information. During times of food depletion, an individual could assess the nutritional status of the population. If food becomes limiting, rodents will begin to catabolise their own muscle protein and the urine will contain larger amounts of protein degradation products. These signals could serve to trigger mechanisms that would affect reproduction. Given that the generation time of rodents is short, complete reproductive inhibition may not be adaptive. However, reduced reproduction may be beneficial. Reduced reproduction would relieve energetic constraints on lactating females that might otherwise jeopardize survival if a full litter size were attempted.

Litters are biased toward producing males when predator or rat catabolic urine is used as a stimulus. This is consistent with theory on reproductive value. Even with reduced litter size, females may still experience lower survival probabilities during reproduction and lactation in food limiting environments because of energetic constraints. However, males would be less constrained by such energetic considerations. Thus, their survivorship probabilities may be higher than females, and by implication their value in contributing to fitness would also be higher. So then, why should rodents reduce reproduction when presented with predator urine? Predators on rodent diets would produce urine with many of the same rodent-derived metabolic products. It is only coincident that the two urines produce the same effect.

The proposed method utilizes naturally derived compounds that pose no environmental hazard. In nature, predators are one of the most powerful extrinsic factors affecting prey population cycles (Hentonnen et al. 1987; Klemola et al. 1997). At the same time, high population density in rodents is the most powerful intrinsic factor for regulation of population density. Our method utilizes combination of intrinsic and extrinsic factors regulating population density under natural conditions. One of the most serious advantages of this method is lack of habituation to repeated exposures to such types of compounds. At the time we keep a sixteenth generation of rats in our laboratory under persistent exposures to predator odors (Voznessenskaya et al., 2006). These animals were still responding to predator urine exposures with reduced litter size. The proposed method should prove useful in reducing our reliance on pesticides with less favorable environmental properties while achieving the goal of reducing rodent populations.

ACKNOWLEDGMENTS

This research was supported by grants from Russian Foundation for Basic Research #07-04-01538a and 10-04-01599a, Russian Academy of Sciences, Program "Zhivaya Priroda" and MK-709.2012.4

KEYWORDS

- **Nontoxic repellency**
- **Population control**
- **Reproduction**
- **Reproductive inhibitors**
- **Rodents**
- **Steroid hormones**

REFERENCES

1. Bacon, S. J.; and Mcclintock, M. K.; *Physiol. Behav.* **1994**, *56*, 359.
2. Bruce, H. M.: *Nature* **1959.** (London), *61*, 157.
3. Dellacorte, C.; Experimental Cell Biology of Taste and Olfaction: Current Techniques and Protocols, Boca Raton: CRC, **1995**, p. 145.
4. Flavelll, S. W.; and Greenberg, M. E; *Annu. Rev. Neurosci.* **2008**, *31*, 563..
5. Hamilton, G. D.; and Bronson, F. H.; *Am. J. Physiol.* **1985**, *250*, 370.
6. Hentonnen, H.; Oksanen, T.; Jortikka, A.; and Haukisalmi, V.; *Oikos*. **1987**. *50*, 353.
7. Huck, U. W.; Pratt, N. C.; Labov, J. B.; and Lisk, R. D.; *J. Reprod. Physiol.* **1988**, *83*, 209.
8. Kassesinova, E.; and Voznessenskaya, V.; *Chem. Senses*. **2009**, *34*(3), E35.
9. Klemola, T.; Koivula, M.; Korpimaki, E.; and Norrdahl, K.; *J. Anim.* **1997**, *66*, 607.
10. Lombardi, J. G.; and Vanderbergh, J. G.; *Science*. **1977**, *196*, 545.
11. Miyazaki, M.; Yamashita, T.; Suzuki, Y.; Soeta, S.; Taira, H.; and Suzuki, A.; *Comp. Biochem. Physiol. B. Biochem. Mol. Biol.* **2006**, *145*, 451.
12. Nolte, D. L.; Mason, J. R.; Epple Aronov, E. V. G.; and Campbell, D. L.; *J. Chem. Ecol.* **1994**, *20*, 1505.
13. Rylnikov, V. A.; Savinetskaya, L. E.; and Voznesenskaya, V. V.; *Soviet J. Ecol.* **1992**, *23*(1), 46.
14. Rutherfurd, K. J.; Rutherfurd, S. M.; Moughan, P. J.; Hendriks, W. H.; *J. Chem. Ecol.* **2002**, *19*, 1405.
15. Sheng, M.; and Greenber, M. E; *Neuron*. **1990** 4, 477.
16. Sokolov, V. E.; Voznessenskaya, V. V.; and Zinkevich, E. P. In: R. L. Doty, D. Muller-Schwarze (Eds): Chemical Signals in Vertebrates 6. **1992**, Plenum Press: New York, 267.
17. Touma, C.; Palme, R.; and Sachser, N.; *Hormones. Behav.* **2004**, *45*, 10.
18. Vanderbergh, J. G.; *J. Endocrinol.* **1969**, 84, 658.
19. Voznessenskaya, V. V.; Krivomazov, G.; Voznesenskaia, A. E. M.; Klyuchnikova, A.; *Chem. Senses*. **2006**, *31*(5), A84.
20. Voznessenskaya, V. V.; and Voznesenskaia, M. A. Klyuchnikova: *Chem. Senses*. 31 (8), E43, (2006).
21. Voznessenskaya, V. V.; Klyuchnikova, M. A.; and Wysocki, C. J.; *Current Zool.* **2010**, 56(6), *813*.
22. Voznessenskaya, V. V.; Naidenko, S. V.; Clark, L.; Pavlov, D. S.; In: Zaikov G. E. (Ed) Biotechnoogy and the Environment Including Biogeotechnology, Nova Science Publishers: NY, **2004**, 59.
23. Voznessenskaya, V. V.; Wysocki, C. J.; Zinkevich, E. P.; In: Doty, R. L., Muller-Schwarze, D. (Eds): Chemical Signals in Vertebrates 6. Plenum Press, New York. **1992**, 281.
24. Whitten, W. K. *J. Endocrinol.* **1956**, *13*, 399.

INDEX